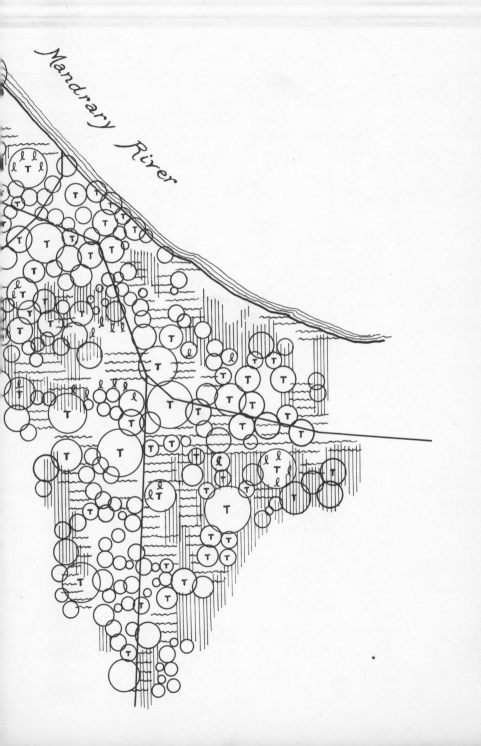

Mandrary River

LEMUR BEHAVIOR
A Madagascar Field Study

LEMUR
Behavior

A Madagascar Field Study

Alison Jolly

The University of Chicago Press
CHICAGO AND LONDON

Library of Congress Catalog Card Number: 66–23690

THE UNIVERSITY OF CHICAGO PRESS, CHICAGO & LONDON
The University of Toronto Press, Toronto 5, Canada

For Pop

Acknowledgments

I should like above all to thank the de Heaulme family, not only for their many kindnesses to me, but for that appreciation of the fauna and flora of Madagascar which led them 20 years ago to set aside and protect the forest reserve at Berenty. I should like also to thank the de Guiteaud family for the opportunity to observe their protected Lemuroidea.

I should like to thank Preston Boggess for his assistance, without which this study could not have even begun; Dr. C. H. Fraser Rowell for his photography of the lemurs; Dr. T. E. Rowell both for her observations and her suggestions about their behavior; Dr. J. and Mrs. V. Buettner-Janusch for intellectual and moral support in the field; and Jeffrey Smith, the most talented camp cook of the lot.

Many people and institutions helped this study in the field. I must first express gratitude to the government of the Malagasy Republic. I am grateful to the Service des Eaux et Forêts, particularly M. Petimaire and M. Ramanansoavina; to the Institut de Recherches Scientifiques, particularly Dr. Malzy, Dr. Roederer, and Georges Randriansolo; to the American Embassy, particularly Mr. Minott and Mr. Allen; to the American Lutheran Mission, particularly Pastor Torvik; to M. and Mme. C. Jenny; and to Mlle. Jeanne Fenistina.

I should like to thank Dr. Capuron, Dr. Brosser, and Dr. H. Moore for identifying the food plants. M. Raucourt of the Institut de Recherches du Coton et des Textiles Exotiques, Mandrary Station, kindly prepared regional climatic charts.

Professor G. E. Hutchinson and Dr. R. J. Andrew read and commented on portions of the manuscript. I should also like to thank all those who helped prepare the manuscript, particularly Alison Mason Kingsbury who drew the maps and Figures 1 and 7.

The New York Zoological Society provided unfailing support. The study was financed by National Science Foundation grant GB–169.

Finally, I should like to thank my husband, Richard Jolly, for much that he did to help, and a few things to hinder, the course of this work.

Contents

I. Introduction   1

II. *Propithecus verreauxi*   20

III. *Lemur catta*   69

List of Maps

List of Figures

List of Plates

List of Tables

Chapter I

Introduction

LEMUROIDEA IN PRIMATE EVOLUTION

Man's emotions have a long history. Sex, aggression, care for the young occur in the lives of all mammals. In primates, these fundamental types of behavior are modified by social organization. Men, apes, Old and New World monkeys, and some prosimians live in permanent groups in which each individual must recognize and adjust its behavior to the character of other individuals. The fascination of studying primate behavior is seeing how social structure varies among different species, and how complex social relations vary among individuals of a species. In our own kind, these relations have reached Proustian subtlety and complication. Still, we cannot but recognize that human relations, and human emotions, stem somehow from the ways of our distant ancestor: the not-quite-monkey who first formed social bonds with others of its kind.

Malagasy lemurs offer a peculiarly exciting glimpse of our history. Their ancestors diverged from ours in the Paleocene or early Eocene, long before the evolution of the New World monkey group (Ceboidea), or of the Old World monkeys which gave rise to apes and men (Cercopithecoidea and Hominoidea). Anatomically, lemurs are in many ways closer to the common ancestor than is any monkey. Yet some lemur species have developed social structures resembling those of monkeys, with more than one male in a troop and with an infant of either sex being likely to remain for life in the troop in which it was born. This kind of social organization, or the behavioral tendencies which produced it, is clearly very widespread, and hence very ancient, among the primates. Lemurs thus provide a third, independent evolutionary line to compare with the New World and Old World lines of social primates. Where the three lines resemble each other, we may find clues to the origin and nature of primate social bonds.

It is not clear when permanent social life began among primates. Many characteristics have been evolving in parallel in the Lemuroidea, Ceboidea, and Cercopithecoidea, including increased binocular vision and increasing brain size (Simons 1962). It is very likely that social life has also developed in parallel in the different lines. New and Old World monkeys may have evolved separately from a prosimian ancestor, lemurs perhaps from an ancestor with a brain smaller than a modern rat's. Some of the modern lemuroid genera are solitary (Petter 1962a); so if they are a homogeneous group, their social organization may have evolved independently on the island of Madagascar.

Simpson (1945) pointed out that it is often very difficult to distinguish between a trait inherited from a common ancestor and a trait evolved in closely related forms under the influence of similar selective pressures. He also showed that we can often consider the two alternatives together, because only animals of close relationship, little distant from their common ancestor, would be likely to develop the same response to similar evolutionary pressures.

Thus, we can assume that the parallel development of three lines of social primates means that the most fundamental aspects of their societies are those found in all three lines and the differences are variations on this fundamental pattern.

Lemurs take on particular importance in this comparison because they retain many primitive characteristics. Their brains are small and narrow compared with the round brain of similarly sized monkeys (Huxley 1861, Elliot-Smith 1902). The olfactory areas of the brain are large. Behaviorally, Lemuroidea mark their territories or each other with individual scents, and *Lemur catta* males challenge each other to ritual stink-fights. The visual (occipital) and "association" areas of the brain are relatively small, which, in fact, goes with their limited understanding of objects—the kind of intelligence measured in psychological tests (Jolly 1964a, b). Lemuroidea breed seasonally, once a year, and have presumably never evolved some higher primates' capacity for year-round breeding. From these three characters alone, the importance of olfaction, the limited intelligence, and the seasonal breeding, we might expect lemuroid societies to resemble those of early, primitive social primates.

But lemurs are also a specialized and peculiar group of animals. Isolated on Madagascar for 50 million years, they have evolved many varied types, from *Microcebus*, the mouse lemur, smallest of living primates, to the spectacular *Indri*, black-and-white and a meter tall, which sings in chorus from hill to hill. There are three separate families of lemurs, ten genera, and about twenty species and forty subspecies (Table I–1). Many more became extinct recently, including *Megaladapis,* a lemur as large as a gorilla. Any group which includes an animal as improbable as the aye-aye should be viewed with suspicion; Malagasy lemurs may be primitive in some characters, but in others they are unique, not to say aberrant.

TABLE I–1

LEMUROIDEA; TAXONOMY [1]

LEMURIDAE	
SUBFAMILY CHEIROGALEINAE:	
Microcebus murinus	mouse lemur
Microcebus coquereli	—
Cheirogaleus major	dwarf lemur
Cheirogaleus medius	—
Cheirogaleus trichotis	—
Phaner furcifer	forked lemur
SUBFAMILY LEMURINAE:	
Lepilemur mustelinus [2]	gentle lemur
Hapalemur griseus	gentle lemur
Hapalemur simus	—
Lemur catta	ringtailed lemur
Lemur macaco [3]	black lemur, brown lemur
Lemur mongoz	mongoz lemur
Lemur variegatus	variegated lemur
Lemur rubriventer	—
INDRIIDAE	
Avahi laniger	woolly lemur
Propithecus verreauxi	sifaka
Propithecus diadema	—
Indri indri	indri
DAUBENTONIIDAE	
Daubentonia madagascariensis	aye-aye

[1] After Hill (1953).
[2] Includes *Lepilemur ruficaudatus* (Petter 1962a, 1965).
[3] Includes *Lemur fulvus* (Petter 1962a, 1965).

This study attempts to describe a few troops of two lemurs (*Propithecus verreauxi* and *Lemur catta*) and to compare them with other primate societies. I have tried to show where they cast light on the evolution of primate social behavior and where they seem to have unique patterns of their own. Hill's invaluable compendium (1953) and the extensive field study of Petter (1962a) and Petter-Rousseaux (1962) provided the background for my own work. The Petters' observations of the solitary Cheirogaleinae, the semisocial *Lepilemur,* and of *Lemur macaco* give a comparative base to discussions of lemur behavior (Table I–2).

The taxonomy followed here is taken from that of Simpson (1945), and the detailed classification of Hill (1953). "Lemuroidea" or the English "lemurs," includes all Malagasy prosimians. I use the term "social Lemuroidea" to include the genera *Hapalemur, Lemur, Propithecus,* and *Indri,* which are always seen in groups. (*Phaner* may well prove, on further study, to be permanently social.)

TABLE I–2

LEMUROIDEA; ACTIVITY, GROUP SIZE, CARE OF YOUNG

	Activity [1]	Group Size [1]	Care of Young [2]
LEMURIDAE			
SUBFAMILY CHEIROGALEINAE:			
Microcebus murinus...............	nocturnal	1–8	nest
Cheirogaleus spp...................	nocturnal	1	nest
Phaner furcifer...................	nocturnal	1–3	?
SUBFAMILY LEMURINAE:			
Lepilemur mustelinus..............	nocturnal	1–2	nest (usually)
Hapalemur griseus...............	diurnal	ca. 6	carried
Lemur catta.....................	diurnal	12–24	carried
Lemur macaco....................	diurnal	4–15	carried
Lemur mongoz...................	diurnal	6–8	carried
Lemur variegatus................	diurnal	2–4	nest
Lemur rubriventer...............	diurnal	4–5	?
INDRIIDAE			
Avahi laniger....................	nocturnal	2–3	carried
Propithecus verreauxi.............	diurnal	2–10	carried
Indri indri.....................	diurnal	3–4	carried
DAUBENTONIIDAE			
Daubentonia madagascariensis......	nocturnal	1	nest

[1] From Petter (1962*a*) and this study.
[2] Petter-Rousseaux (1964).

In one respect I depart from accepted nomenclature and follow Petter (1962*a*, 1965) by including *Lemur fulvus* with *L. macaco.* The *L. macaco* group is very confusing, with several different chromosome numbers even in the single subspecies *L. macaco fulvus* (Table I–3). However, there are few behavioral differences. The vocalizations of *L. macaco macaco* can be distinguished by ear from those of other subspecies, but this is also true between subspecies of *Propithecus verreauxi.* Pending taxonomic revision of the genus, it is most convenient for a student of behavior to group observations of *L. macaco* subspecies together and then contrast them with the sharply different *L. catta* and *L. variegatus.*

The suggestion of Bolwig (1960) that *L. catta* be called "Odor-lemur" seems unnecessary. Hill (1953) concludes that *L. catta* deserves at least subgeneric separation from other *Lemur.* Hill points out, however, that if *L. catta* is generically distinct, its name must be not "Odorlemur," but *Lemur,* for it takes precedence as the original lemur described by Linnaeus (1758).

LITERATURE

Greek and Arab geographers knew of Madagascar, which the Arabs romantically called "al Qumar," the "island of the moon" (Kent 1962). Marco Polo (1295) first wrote the name Madeigascar and spoke of the roc (possibly *Aepyornis*) which lived there, but he said it preyed on elephants, which certainly were not there.

The Sieur Etienne de Flacourt described the fauna in 1661. He gave identifiable accounts of *Indri, Hapalemur, L. macaco fulvus* or *albifrons, P. verreauxi verreauxi, L. catta,* and a "squirrel" which may be *Lepilemur.* Although he devoted only a sentence or two to each one, all his statements are correct. *P. verreauxi verreauxi* "was not again noticed," in Hill's (1953) phrase, for two centuries. This is surprising, since Flacourt recorded an unequivocal description, the correct Malagasy name "sifac," and a location where they still exist (in fact, the area where I studied). It is even conceivable that he saw one of the now extinct Indriidae. After all these accurate details, Flacourt claimed to have himself glimpsed the "Tratratratra"—a quadrupedal animal as large as a 2-year-old calf with monkey's hands and feet and human face and ears.

There followed a long period of description and cataloguing, well summarized by Hill (1953). A *L. catta* was apparently the first lemur to reach Europe alive. It was drawn, with other lemurs, by Edwards (1751, 1758). The discovering, classifying, and dissection of new species continued through the publication, in 1876 and from 1890 to 1896, of the "Histoire de Madagascar-Mammalia" by Milne-Edwards and Grandidier. This included color plates of indriid and lemurid face-masks showing variants of and intermediates between many modern recognized forms. Hill's (1953) comprehensive monograph summarizes most of the taxonomic and anatomical knowledge to date.

Meanwhile, many authors have pointed out the place of prosimians in the scale of primate evolution. Linnaeus recognized *L. catta* as worthy of a distinct suborder of primates as early as 1758. In 1863 T. H. Huxley made his celebrated statement, "Perhaps no order of mammals presents us with so extraordinary a series of gradations as

TABLE I-3

LEMUROIDEA; TAXONOMIC LIST,[1] CHROMOSOME NUMBERS,[2] DISTRIBUTIONS [3,4]

	Chromosomes (2n)	Eastern Region				Sambirano		Western Region				Southern Region		
		1	2	3	4	5	6	7	8	9	10	11	12	13
LEMURIDAE														
Microcebus murinus murinus	66	H	PH			H		P	JH	PH	JH	PH	JH	
Microcebus murinus smithii (=rufus)				H	PH			H						
Microcebus coquereli						H	H							
Cheirogaleus major	66	H	P			H	H	H	H	H	H	H	H	H
Cheirogaleus medius								P	H	P	J			
Phaner furcifer					PH		PH	H	J	P	J			
Lepilemur mustelinus mustelinus				PH										
Lepilemur mustelinus microdon		H	P											
Lepilemur mustelinus leucopus												H	PJH	JH
Lepilemur mustelinus ruficaudatus								PH	PH	PH	PH			
Lepilemur mustelinus dorsalis					P	PJH	H	H	H	H	H	H	H	H
Hapalemur griseus griseus	54	J	PJ											
Hapalemur griseus olivaceus	58	H	H	H	H									
Hapalemur simus			P		P									
Lemur variegatus variegatus	46			H										
Lemur variegatus ruber	46													
Lemur variegatus subcinctus			H	H										
Lemur macaco macaco	44					PJH	PJH	H						
Lemur macaco flavifrons							H							
Lemur macaco sanfordi	60													
Lemur macaco albifrons	48		PJH	PH	PH									
Lemur macaco fulvus (eastern)	60													
Lemur macaco fulvus (western)								PJ						
Lemur macaco collaris	48,54	JH							JH					
Lemur macaco rufus	60							H						
Lemur macaco mayottensis														

TABLE I-3 (continued) LEMUROIDEA

	Chromosomes (2n)	Eastern Region				Sambirano			Western Region			Southern Region		
		1	2	3	4	5	6	7	8	9	10	11	12	13
Lemur mongoz mongoz	60							PJ						
Lemur mongoz coronatus							H	H						
Lemur catta	56	H							H	PH	PJH	H	PJH	JH
Lemur rubriventer			PH	H	H	H	H							
INDRIIDAE														
Indri indri			PJH	PH	PH									
Propithecus verreauxi verreauxi	48											PH	PJH	JH
Propithecus verreauxi melanotic, (-majori)									JH	PH	PJ			
Propithecus verreauxi deckenii														
Propithecus verreauxi coronatus	48							PJH						
Propithecus verreauxi coquereli	48													
Propithecus diadema diadema			PJH	PH										
Propithecus diadema edwardsi														
Propithecus diadema holomelas					PH									
Propithecus diadema candidus				PH	H									
Propithecus diadema perrieri							H	PH						
Avahi laniger laniger		JH	PH	PH	PH									
Avahi laniger occidentalis								PH						
DAUBENTONIIDAE														
Daubentonia madagascariensis		H		PH	PH		H							

[1] After Hill (1953) and Petter (1962a).
[2] Chu and Bender (1962), Buettner–Janusch (1962).
[3] Localities 1–13 shown on map II.
[4] Hill (1953) = H.
Petter (1962a, c) = P.
Personal observation = J.

Distributions from Hill are speculative syntheses of earlier work, while Petter and Jolly give direct observations. Petter attempted to see all species present in an area; so his negative reports are of value. Jolly did not; so only her positive reports are of value. Petter visited all areas except 1, 8, and 13.

this—leading us insensibly from the crown and summit of the animal creation down to creatures, from which there is but a step, as it seems, to the lowest, smallest, and least intelligent of the placental mammals." He also said (1861), "the lemur's cortex is so little developed it leaves the cerebellum exposed in dorsal view—by this fact alone the lemur is farther removed from the monkey than the monkey is from man." Elliot-Smith (1902) analyzed the lemuroid brain in detail and showed that the olfactory region of the brain was large, and the visual area small, compared with those of higher primate brains. He argued (1927) that prosimians represent a transition stage from dependence on smell to dependence on sight—a halfway step to the monkey with its attention to visual detail of the environment. Clark (1959) has extended this approach to prosimian anatomy and its relation to behavior.

There have been a number of recent studies of lemuroid behavior in captivity. Andrew (1963a, 1964a) comprehensively described the displays of captive prosimians in relation to ethological concepts and in relation to their probable social organization. I studied captive prosimians' use of their hands, their manipulation of unfamiliar objects, and their learning of simple object problems (Bishop 1962, 1964; Jolly 1964a, b). Starnmühlner (1960) described the general behavior of *Microcebus*, Buettner-Janusch and Andrew (1962) published a note on grooming behavior, and Petter-Rousseaux (1962, 1964) extensively described reproductive and maternal behavior in both wild and captive lemurs.

The chief early account of habits of wild lemuroids is Shaw's (1879). He described *L. catta* running over bare rocks with no trees for refuge, which may be possible but is hardly typical for the species. Rand (1935) added ecological observations of *Lemur* and *Propithecus*. Millot (1952) gives accounts of the explorative curiosity of *Lemur*, with a photograph of troops of *L. macaco* and *L. catta* simultaneously investigating pirogues by a river. Bolwig (1960) made notes on the behavior of *L. catta* in the wild as well as describing many of its displays in captivity. Attenborough (1961) was the first of the current lemur watchers and visited many different habitats.

The Petters have made the only general survey of the behavior of wild Malagasy lemurs as well as describing prosimians' reproductive behavior. (Petter 1962a, Petter-Rousseaux 1962). My own study depends heavily on their work. In fact, Petter's extensive field study with descriptions of different species, areas, and seasons is the necessary complement to the present, highly localized, intensive account.

Finally, Buettner-Janusch has sponsored a number of recent

field investigations as background to his work in primate blood components. (Buettner-Janusch 1962, Buettner-Janusch and Buettner-Janusch 1964, Buettner-Janusch, 1966). Observations by his students, Maxim, Boggess, and Smith, are quoted as invaluable comparative data.

SCOPE AND METHOD

SPECIES AND PERIODS OF OBSERVATION

This study describes the behavior of two social Lemuroidea, *P. verreauxi verreauxi* and *Lemur catta*. I watched them chiefly in the gallery forest at Berenty, on the banks of the Mandrary River, in Madagascar. I also observed these same lemurs nearby at Bevala in modified gallery forest. Observations in the forest of Lambomakandro, in southwestern Madagascar, allowed comparison with a different habitat at the other end of these lemurs' range. I am much indebted to the 5-week field study of Boggess and Smith (Buettner-Janusch, 1966) at Lambomakandro for comparative data. Brief visits to other forests proved illuminating, for in a group as diversified as the Lemuroidea, it is important to see the differences between subspecies, species, and genera.

In the Mandrary region, I watched *P. verreauxi* for 255 hours and *L. catta* for 400 hours. Of this, 80 hours was observation of both species at once. I worked in the Mandrary region from February 11 to April 6, May 11 to May 14, May 28 to July 18, July 25 to July 30, and August 13 to September 5, 1963 and March 17 to May 1, 1964. No observations were made there from October through January.

I also visited the Eastern rain forest domain at Perinet from November 19 to November 20, 1962 and from May 1 to May 6, 1963, and at Bemangidy, from July 19 to July 21, 1963. In the Sambirano domain I spent from December 9 to December 19, 1962, at Nosy-bé and Ambanja, and in the Western domain from November 28 to December 6, 1962, at Ankarafantsika, and from January 18 to January 19 and April 9 to April 16, 1963, at Lambomakandro. The species I saw on these trips were observed from 20 minutes to 8 hours, with two to forty encounters.

Table I–3 lists the varieties of Lemuroidea, with their distributions. Hill (1953) has summarized the earlier literature, much of which is vague about distribution. Petter (1962b, c) made an extensive survey of lemuroid species and gave precise locations. Petter probably saw all the species in each area he studied, which I did not attempt to do. Table I–3 also shows chromosome numbers when they are known from the work of Chu and Bender (1962). It is clear that recent work does not solve the riddles of lemuroid speciation; the problem remains as complex as before.

LOCATION (MAPS I–IV)

Madagascar is about 1500 km long, stretching from 12° to 26° S. Lat., its southern tip well below the Tropic of Capricorn (Map I). Its central highlands lie 1800 m above sea level, so that it has both north-south and altitudinal gradients of temperature. The prevailing winds come from the east; so there is also a rainfall gradient from east to west. As a result, the plants are sharply divided into phytogeographic zones (Maps II, III). The wet oriental forest is distinct in almost all species from the dry occidental forest. The oriental zone is tangled, tree-ferned jungle; the occidental zone bears open woodland. In most of the island, the two forests never meet, for the central plateau is burned over, bare, and too eroded to support Madagascar's native flora. In the south, however, the dry and wet regions come into dramatic opposition (Map IV).

As you stand on the western edge of the Mandrary River Basin, you look eastward over gray euphorbia bush, over the narrow strip of green gallery forest and the glint of river and red sandbank, toward a range of mountains 2000 m high between the desert and the eastern sea. At this distance the mountains are featureless blue, their volcanic skyline hidden by a whipped cream topping of cloud. The Indian Ocean lies just beyond to the east, but the clouds stop as though cut off with a knife along the mountains' western slope. Two hours' drive eastward (70 km) will carry you from desert to mountain rain forest, from rainfall of 50 cm/year to one of 300 cm/year.

The Mandrary River flows not only through a dry region of Madagascar but also through one of the hottest. Average daily temperature in 1963 ranged from 9° to 33° C, with a high of 43° C. The warmest months are from December to February, the coldest July and August. The hot season is also the wet season, although total rainfall varies greatly from year to year. From December 1962 to February 1963, 48.5 cm of rain fell, but in the same months in 1963 and 1964, only 17.5 cm fell. Between July and September 1963, total precipitation was less than 2 cm (data from I.R.C.T. Mandrary Station).

The red desert sand supports one of the strangest forests in the world. The "forêt épineuse" or "bush" is dominated by spiny euphorbia, and more than 80 per cent of the plant species are endemic. Ten- to 13-m *Didiera madagascariensis* recall the candelabra cactus of America's southwest. Their woody gray trunks tower up and separate into a bunch of vertical or horizontal arms. Helical rows of minute

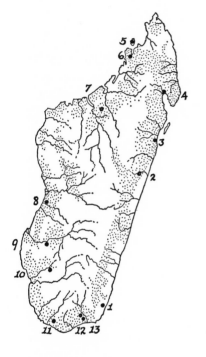

I
RIVERS AND FORESTS

STUDY SITES •

Eastern Domain
1 Bemangidy: Jolly
2 Perinet: Petter, Jolly, Boggess, Smith
3 Fenerive: Petter
4 Maroansetra: Petter

Sambirano Domain
5 Nosy-bé: Petter, Jolly
6 Ambanja: Petter, Jolly

Western Domain
7 Ankarafantsika: Petter, Jolly
8 Morondava: Jolly
9 Mangoky: Petter
10 Lambomakandro:
        Petter, Jolly, Boggess, Smith

Southern Domain
11 Ivohibé: Petter
12 Ifotaka: Petter, Jolly
13 Bevala, Berenty: Jolly

MAP I.—Remaining areas of indigenous forest are stippled. The marked sites have been visited during recent studies by Petter (1962*a*, 1962*c*), Boggess and Smith (Buettner-Janusch, 1966), and myself. Site 13 includes the Reserve where I chiefly worked.

leaves, set vertically to escape the noon sun, coil up the trunk, each two leaves separated by a thorn. Between the *Didiera* crowd stunted acacia thorns, aloes, fleshy euphorb "latex trees," and succulent vines. The plants grow close together, their long spines and fleshy fingers interlaced. Each plant however, has reduced its leaf area to a minimum; so there is no shade in the tangle of branches.

Forty years ago, it was almost impossible for a person to push through this weird forest because *Opuntia monacantha,* the prickly pear with its 10-cm spines, grew everywhere. These plants were introduced in 1769 from Mexico via Réunion (Decary 1947). They made fortress walls for Tandroy villages and a reserve of succulent vegetation for cattle and other animals in the dry season. The *Opuntia* preserved the Antandroy from conquest by the Imerina first, and later by the French. The French succeeded in controlling the area only by negotiation with local chiefs, who revealed paths through the forest. In 1924, however, a still unidentified botanist introduced the cochineal bug, *Dactylopius coccus* (Humbert 1947). The insect swarmed over the area and wiped out the *Opuntia,* incidentally causing many Antandroy cattle to die of thirst and their masters of famine. Now cattle are herded for 20 or 30 km every 2 days for water, often to the Mandrary River. There is no way of telling how much the ecology of the spiny forest was changed by the appearance and disappearance of the *Opuntia,* or whether the forest has returned to its original condition.

Animals that survive in the euphorbia forest, including *P. verreauxi verreauxi,* are those that can find shade in the bundle of a *Didiera's* stems and get water from its leathery leaves. There is presumably a strong, constant physiological pressure on the desert animals to conserve water during the dry season. There is also probably great competition among them for shade, especially holes to hide in; even the nocturnal lemurs, *Microcebus murinus murinus* and *Lepilemur mustelinus leucopus,* and hole-nesting birds like the lovebird *Agapornis cana* must compete. A bird's egg in an unshaded nest would be hard-boiled soon after sunrise.

The Mandrary itself (Plates 1 and 2) forms a green and unexpected oasis in the desert. The river drains an area of nearly 10,000 km ^{2}, including tributaries from the wet eastern mountains. It floods during the rains; some years it will dry out by September or October. Then the Antandroy with their oxcarts dig 3–5 m into the sand of the riverbed until they find water. The stream meanders over a bed perhaps 100 m wide, and it slices between 10 and 15 m from the outside of a curve during the rains. It leaves old river arms and flood

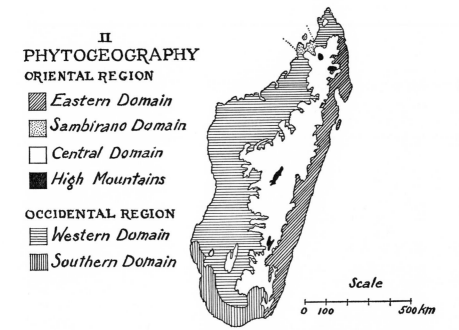

II
PHYTOGEOGRAPHY
ORIENTAL REGION

Eastern Domain

Sambirano Domain

Central Domain

High Mountains

OCCIDENTAL REGION

Western Domain

Southern Domain

Scale

0 100   500 km

MAP II.—Vegetation zones defined by Humbert (1955). They result directly from the relief and rainfall gradients shown in maps III and IV. The Central Domain has almost completely lost its indigenous woodland (Map I), while most other types of wood or forest are still represented. The main study area (Site 13, Map I) lies in the Southern Domain of euphorbia bush.

plains filled with gray alluvial clay at a level several meters below the surrounding red sand.

Huge *Tamarindus indica* trees, locally called "kily," dominate the gallery forest. The kily was perhaps introduced to Madagascar by early Arab traders, who brewed the seeds for a laxative. Kilys have barrel trunks, low heavy branches, and green round tops like a white oak, with the compound leaf and bean pod of a legume. The gallery forest gives the impression of being a 500-year-old oak woods—indeed it is a temperate zone wood a hundred miles south of the Tropic of Capricorn. Between the kilys rise tall feathery acacias and *Albizzias* and the ubiquitous *Celtis bifida*, the "silikata sifaka," or "tree–*Propithecus*—cannot-climb.*" But the kilys dominate, in the same way one species may dominate a north temperate woodland. Map V is a diagram of vegetation types. The wood as a whole is about 21 to 30 m high, with more or less continuous but not very dense foliage from 13 to 21 m; the kilys and other tall trees protrude to 30 m. Saplings make up most of the undergrowth, with bramble-like bushes in more open spaces. Open grassy meadows and clearings interrupt the woods. The only obvious tropical feature is banks of glossy-leaved lianas that line the clearings and cascade from the tops of high trees in an impenetrable green curtain.

The gallery forest makes a narrow strip almost 80 km long on the lower Mandrary and its tributary, the Mananary, wherever there is alluvial soil. A sharp 1- to 7-m bank of red sand bounds its outer edge, with the spiny desert on top. It is often possible to walk from one climax to the other, from a kily to a *Didiera* in ten strides, or, for a lemur, one leap.

The reserve where I worked consists of about 100 hectares of gallery forest (250 acres) and half a hectare of *Didiera* forest, although part of the gallery forest (perhaps 30 hectares) is somewhat degraded. The whole is surrounded by sisal fields, so that it forms an island of natural habitat isolated by terrain which the wildlife cannot cross. The fauna of the surrounding area took refuge there, and the reserve itself has existed for 20 years; so the fauna are probably approaching equilibrium at the highest population density the reserve can support, unless some species' populations have gone into cycles.

The 10 hectares I studied most intensively (Map V) held the home ranges of four *P. verreauxi* troops and parts of the ranges of five others, a total of forty-two animals (before the birth of 1963 infants). The same area contained the home range of one *L. catta* troop and parts of the ranges of two others, a total of forty-eight animals. Judging by the numbers in this area I believe there might be

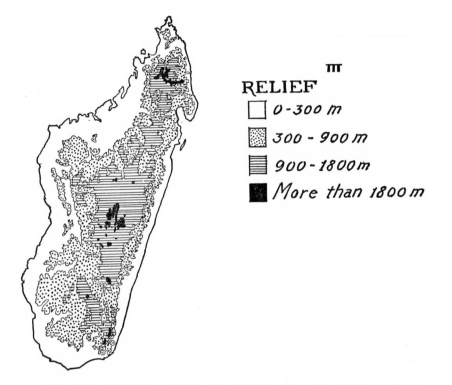

Map III.—Relief after the handbook of the Commissariat Generale au Plan (1962). Steeply rising eastern mountains break the prevailing winds, causing an east-west rainfall gradient (Map IV). Marked temperature gradients exist between the hot coastal areas and the high, cold central plateau. In the southeast corner of the island lies a large cirque of mountains, originally a single volcanic crater, which surround the Mandrary watershed and the main study area (Site 13, Map I).

IV
RAINFALL

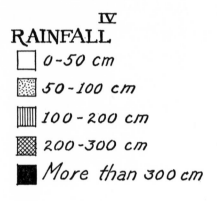

- ☐ 0-50 cm
- ▦ 50-100 cm
- ▥ 100-200 cm
- ▨ 200-300 cm
- ■ More than 300 cm

MAP IV.—Precipitation zones after the handbook of the Commissariat Generale au Plan (1962). The main study area (Site 13, Map I) fell in the region of 50–100 cm annual rainfall. Note the abrupt transition just eastward of the site to mountain rainforest (maps I, II, and III).

STUDY AREA

V

SCALE: |———| 100m

——— PATHS

Ⓣ TAMARIND TREES

◯ OTHER LARGE TREES

|||||| TREES LESS THAN 13M TALL

∿∿∿ BUSHES

ℓ ℓ ℓ LIANAS

Blank BARE GROUND OR MEADOW

MAP V.—Main study area, a portion of the Berenty Reserve. Map constructed from compass bearings and paced distances. This map shows the general conformation of the gallery forest, with a broken canopy of large trees 13–30 m tall. Beneath the large trees were saplings or bare ground, while in clearings where light penetrated were bushes or meadow. Lianas hung from the large trees on the edge of open clearings.

a maximum of about fifty *P. verreauxi* troops, or two hundred animals, and about twenty *L. catta* troops, or three hundred fifty animals, in the whole reserve. The actual totals are somewhat lower, since not all the forest is equally well suited to lemuroids.

I also visited the government reserves of Ankarafantsika in the Western domain, Lokobe, on the island of Nosy-bé, in the Sambirano domain, and Perinet, in the Eastern domain. These forests are described in Petter (1962a), and Humbert (1955). Bémangidy also lies in the eastern forest zone.

The forest of Lambomakandro is described by Petter (1962a). It is particularly interesting because *P. verreauxi verreauxi* and *L. catta* live there in association with *L. macaco rufus*. It is a Western woodland of temperate type, though slightly north of the Tropic of Capricorn. Like the gallery forest at Berenty, it has continuous, moderately dense foliage from 13 to 21 m, with tall trees up to 30 m and the foliage of smaller trees at 7 to 13 m. Undergrowth again is bushes and brambles, mostly in areas where more light penetrates. The 15 hectares I studied for 2 weeks and Boggess and Smith (Buettner-Janusch, 1966) studied for 5 weeks lie in a valley near a swamp. The hills on either side of this valley have much drier vegetation, including thorn trees, aloes, and succulents, although I never found the climax euphorbia forest of the Southern desert.

PROCEDURE

I observed the lemurs with 7 x 50 binoculars and took notes of their behavior. I used no blinds, although I tried to sit partly screened by a low bush and to move as little as possible.

Lemurs which are quite unaccustomed to man will not flee until he is within 10 to 15 m of them. Most accessible wild lemurs, however, have been hunted, and so treat man as a potential enemy and disappear if he approaches. Animals that have been long protected and have seen people before approach within 5 to 8 m of a stranger, with "ground predator" or "mobbing" calls. The wild lemuroids at Berenty eventually became tame enough to carry on normal activities 3 m from me, glancing at me only if I moved suddenly. On one occasion two juvenile *L. catta* approached Dr. C. H. Fraser Rowell and bit his shoelaces. Usually, however, they ignored me, as they would another species of lemuroid. They were, however, almost always aware of my presence. On the rare occasions when the animals had not seen me, this was quite clear both by their lack of glances in my direction and their start when they noticed me.

The animals were never "provisionized," not even incidentally from my lunch.

Still photographs were taken with an Alpa 35-mm camera, with 50; 180-, and 350-mm lenses. Movies were made with a Bolex H16 reflex with 25; 75-, and 150-mm lenses. Vocalizations were recorded on an Amplifier Corp. tape recorder. Usually, however, I carried only a notebook and binoculars during observations.

Chapter II

Propithecus verreauxi verreauxi

GENERAL DESCRIPTION

Propithecus are Indriidae in the family of Indri and Avahi. The two species, *P. diadema* and *P. verreauxi* occur, respectively, in the eastern and western zones of Madagascar. Each is divided into subspecies, four for *P. verreauxi* and five for *P. diadema,* which are distinguished chiefly by the color of their fur although many intermediate forms are known (Hill 1953). The races are sharply localized, often separated by large rivers. *P.v. verreauxi,* the true "sifaka," occupies all the dry south and southwest.

P. v. verreauxi is one of the large, more spectacular lemurs (Plate 3). Its head and body measure 45 cm (Hill 1953), but standing on its hind legs, it is more than a meter tall. It has deep, white fur and black skin, which is bare on palms, soles, genitalia, and on the black heart-shaped face. A seal-brown cap starts above the white brow and covers the back of the animal's head and neck. As a *Propithecus* watches you from a vertical trunk, you are first aware of a gleaming white body, a square, white-furred head, and the black heart-shaped face drawn upon it. Two yellow eyes stare directly toward you over the top of a muzzle. A slim tail hangs down behind or curls up on itself like a mainspring. Then, without warning, the *Propithecus* leaps. It seems to double in size. The hind legs, longer than the head and body, spring open to propel the animal backward or sideways as much as 30 feet in a curve taut as ballet. The *Propithecus* turns or half-turns in mid-air to soar with both hind feet first. The legs break its leap and fold, and it again clings vertically to a trunk, still watching you.

This general impression gives way to recognition of age, sex, and individual differences. *Propithecus* has no sexual dimorphism in color. Females may be slightly smaller than males, on the average, but in any troop a female may appear larger than a male. In the wild, apparent size depends so much on the condition of the animal's fur, that young bushy males appear larger than ragged old ones and healthy females appear to be between the sizes of the two.

Males have a distinct penis, especially visible during urine-marking. The female genitalia vary, but generally females have a small clitoris, although the labia may be so large and black that they resemble a scrotum. The female may show a pale, unpigmented vaginal opening throughout the year. During estrus the opening enlarges and flushes deep reddish-pink.

Nipples are only visible in a nursing female, on whom one pair can be seen high up near the axillae. In the field, scent glands are more useful secondary sexual characters. Adult male *Propithecus* have a long scent gland down the ventral surface of their throats. Petter (1962a) discovered this by watching their characteristic gesture of throat-marking. Males also urine-mark with the tip of the penis (Petter 1962a) and very rarely rub the anogenital region on a branch as the females do.

In practice it is easiest to determine sex from the genitalia or throat-marking at the same time you are observing the individual characteristics of an animal; thereafter you can recognize the animal by its face.

Age characteristics are easier to judge. Infants are born in early July in the Mandrary region and at about the same time in *P. v. coquereli* in the Ankarafantsika forest at the opposite end of Madagascar (Petter 1962a). A mother carries her young on her belly until October. Then the young rides on her back until December or January, when it is half to two-thirds grown. The 1-year-old juvenile is still a little smaller than an adult when the next year's infants are born. *Propithecus* may be able to breed in its second year of life, but it also may not be, since *L. catta* probably cannot. By March, the 21-month animals are indistinguishable from adults.

Adults usually have glossy, thick fur and well-rounded outlines. However, many troops contain old males, which are thin and shrunken with balding thighs and gray or reddish caps. Many adult males show nicks in their ears from fighting. The old males have very tattered ears, which shows they are scarred veterans, not merely sick animals.

Individual differences are relatively clear (Fig. 1). Males can be distinguished by the number and shape of nocks in their ears (Petter 1962a), although not all males have these nocks.

There were few visible injuries, except for bitten ears. Four males had some of the fur scraped off their tails, leaving tufts or

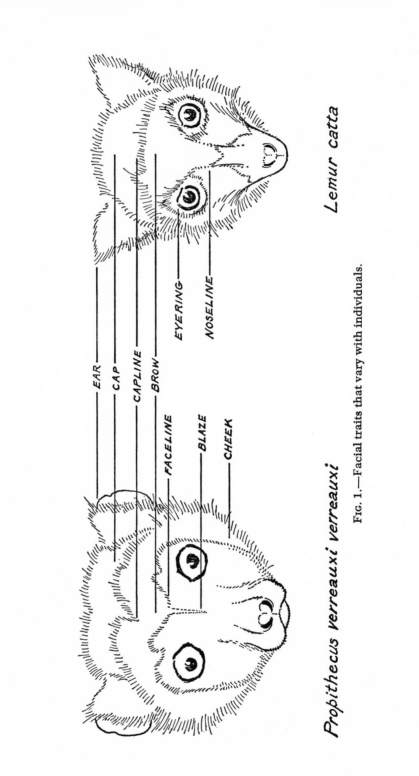

Propithecus verreauxi verreauxi

Lemur catta

Fig. 1.—Facial traits that vary with individuals.

tassels on the end. One female had a broken tail whose end dangled downward. One adult male had only one eye, the other being a shrunken socket. A juvenile was once seen with its eye running and stuck closed; it kept wiping the eye with the radial side of its wrist. Two days later, however, the eye seemed normal.

Facial patterns differ in each animal, particularly the shape of the white browband between the hairless face and brown cap. The capline may be square, jagged, or peaked up or down, often assymmetrically, and so may be the faceline. A few animals have distinct brown patches on the cheeks or a white blaze down the nose. The cap itself varies from chocolate to caramel color and often in old males takes a bleached gray or reddish tint.

Normal, white *P.v.verreauxi* may also have white, blond, or caramel colored chest fur and a grayish patch of varying shade on the lower back. A few have pale orange shoulders and thighs, with darker gray backs, which approaches the coloring of *P. diadema diadema,* that is, bright orange limbs and a charcoal gray back.

The so-called *P.v.majori* (ROTHSCHILD 1894) (Hill 1953, Petter 1962*a*) is simply a melanotic form of *P.v.verreauxi,* with dark-brown sideburns and chest and brown "eagle wing" patterns on forearms and thighs. The maroon markings of *P.v.coquereli* are similar. Melanotics vary in depth and distribution of color, just as normals do, but there are few or no real intermediates between the two forms. I have observed three mixed troops of white and melanotic animals in two areas and seen one melanotic animal in a third area; it seems clear that the melanotics are within the usual range of individual variation of *P.v.verreauxi* (see pp. 175–177).

In the Reserve, I named *Propithecus* troops after their most easily identified individual. The maps and tables of this chapter refer, for instance, to the BN troop or to the DE troop, that is, the Blaze Nose or Dent Ear troop. The individual named Blazenose was a large female whose 1963 and 1964 infants both inherited her unusual facial pattern, while Dentear was a pugnacious male with most of his left ear bitten away.

ECOLOGY

HOME RANGE AND TERRITORY

The Berenty Reserve was established 20 years ago. Many animals took refuge there as neighboring land was put under cultivation. It is quite likely that the reserve is supporting nearly the maximum possible density of lemurs, unless the populations fluctuate cyclically.

Propithecus, like other lemurs, is highly territorial. In the reserve, territorial size was apparently limited by the pressure of neighboring groups. Petter (1962a) observed that *Propithecus* territories vary in size, being large where the forest is sparse and small where it is thick. It is reasonable to assume that at high population densities, the range and territory of a group represent some minimum combination of behavioral-ecological requirements.

A group's home range is the total area where they are found; its territory for the purposes of this discussion is the area other groups did not enter.

Territories and home ranges of four *P.v.verreauxi* troops in the reserve, shown in Map VI, were fairly constant throughout the seasons observed. Territorial disputes were frequent; whenever two *P.verreauxi* groups met, one chased the other. The ranges of different groups overlapped, but each group had a nucleus of territory that others did not penetrate. Map VI also shows observed territorial disputes and which group won.

In 1963 home ranges of four troops (each with five adult animals), the Widow Peaks, Dent Ears, Blaze Noses, and Red Heads, were about the same size, from 2.2 to 2.6 hectares (Table II–1). Their territories varied more, from 0.6 to 1.8 hectares. The three troops which had territories of 1.2 to 1.8 hectares remained unchanged from March, 1963, to April, 1964, while the Red Heads, with only 0.6 hectares, split up. Two of the Red Head troop members joined a solitary male, Sideburn, who had a range of 1 hectare but no territory of his own.

The troops had no distinct "core areas" (DeVore and Hall 1965), except for remaining in certain vegetation types.

Map V shows the major types of vegetation in the gallery forest, as they are used by the lemuroids. Circles are trees taller than 13 m. The kilys (*Tamarindus*) are marked *T*.

Propithecus moves freely in any large tree and sleeps and eats there. The area of large trees in the home ranges was nearly constant, from 1.2 to 1.5 hectares. In the nuclear territories, the three stable troops had 0.9, 0.8, and 0.8 hectares, respectively; the Redheads could defend only 0.2 hectares (Table II–1).

Vertical hatching in Map V indicates trees and saplings less than 13 m in height. The *Propithecus* fed in their foliage or hopped between their vertical trunks at ground level or even descended them to the ground. Similarly, *Propithecus* may hop between, or climb down, the trunks of lianas, marked *l* on the map. The small trees and lianas compose the lower layer of habitat available to *Propithecus*

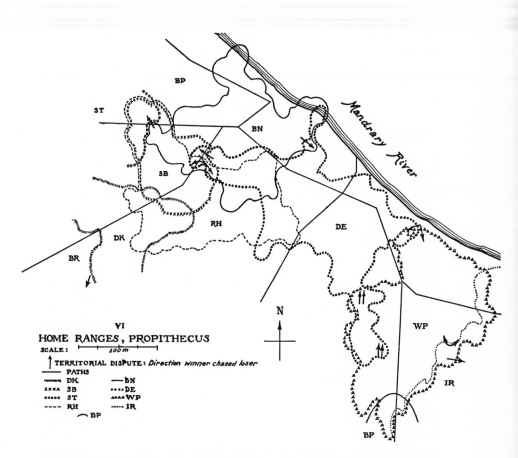

MAP VI.—*Propithecus* troop home range boundaries. Area the same as Map V. Territorial battles took place in zones of overlap. Troop territory defined as area other troops did not enter.

and are important for food, for pathways, and for shade for rest at midday in the hot season.

Area of this available layer varied, however, among territories. There were 0.5 to 1.2 hectares of small trees and lianas in the troop ranges and from 0.2 to 0.5 hectares in their territories, which is relatively a greater variation than in the area of large trees (Table II–1).

TABLE II–1

Propithecus verreauxi HOME RANGE AND TERRITORY, 1963
(areas in hectares [1,2])

Troop	WP	DE	BN	RH	SB
TROOP SIZE	5	5	5	5	1
HOME RANGE					
Total	2.6	2.6	2.2	2.2	1.0
Large trees	1.3	1.5	1.5	1.2	.4
Small trees	.5	1.2	.6	1.2	.2
Bushes.................	.5	.3	.3	.3	.3
Open ground...........	1.5	1.2	1.3	.7	.5
TERRITORY [3]					
Total	1.8	1.5	1.2	.6	0
Large trees	.9	.8	.8	.2	–
Small trees	.3	.5	.2	.5	–
Bushes.................	.5	.2	.2	.1	–
Open ground...........	1.0	.8	.8	0	–

[1] Areas measured from map V.
[2] 1 hectare = 2.47 acres = 0.01 km².
[3] Territory refers to area which was the sole property of a troop, which other troops did not enter, not to the entire area a troop attempted to defend.

Horizontal scalloped lines (Map V) show clumps of bushes. Bushes can be distinguished from small trees because small trees have at least 5 feet of clear trunk below their foliage and bushes grow down to the ground. In bushes *Propithecus* cannot rapidly hop between stems; so they did not venture into clumps of bushes. Bushes thus form barriers around which *Propithecus* must detour. White areas are meadow or bare ground. Under the kily trees, fallen leaves and pods blanket the ground, and there is no ground cover. In the open grow grasses and low Acanthids. Although *Propithecus* are able to cross open spaces on the ground, they rarely did so. In general, *Propithecus* only descend to the ground when lianas or small trees offer immediate escape routes. Areas of bushes and open ground are thus unused parts of the territory.

Map V was made at the end of the observation period of 1963, when I knew which general types of vegetation were significant for the lemurs; therefore I do not have records of the amount of time the animals spent in each type. Table II–2 gives a rough picture, since all the time spent above 13 m was by definition in large trees, time below 7 m was in or under small trees, with overlap in the 7- to 13-m range. This confirms the impression that *Propithecus'* vertical and horizontal distribution in this sort of forest depends chiefly on the large trees.

TABLE II–2

Propithecus verreauxi; HEIGHT FROM THE GROUND

METERS	PER CENT OBSERVATIONS			
	Mar.–Apr., 1963	June–Sept., 1963	Mar.–Apr., 1964	Avg
0........	5	5	5	5
>0–3........	5	5	5	5
>3–7........	20	10	20	15
>7–13.......	35	25	15	25
>13–21.......	30	40	50	40
>21........	5	15	5	10

A troop of *P.verreauxi* may travel across its nuclear territory twice a day, although it usually does not go as far as the limits of its home range. In fact, a troop could move from one end of the range to the other in 5 minutes, if it maintained speed. However, even when the troop is fleeing or moving from one area to another, 100 m in 15 minutes is a major displacement.

The troop may follow the same route through its home range for 3 or 4 successive days but then shift abruptly to another part of the range, with different sleeping and feeding trees. In this way the troop visits all parts of the range every week or 10 days, and damage to food plants is evenly spread over the entire area.

BRANCH TYPE

Petter (1962*a*) asserts that the vertical posture of *Lepilemur* and the Indriidae, including *Propithecus*, sets them apart from more generalized quadrupedal Lemuroidea. He believes that the Indriidae are confined to vertical branches and trunks and handicapped on small or horizontal branches. They would thus be separated ecologically from quadrupedal *Lemur* such as *L.catta*, with which they share the forest.

This is true to some extent. *Lemur* species, including *L.catta*, do

not cling and rest on vertical trunks as *Propithecus* does, although
Lemur often leap on such trunks while moving through the forest.
When a *Propithecus* troop is moving, however, the animals hop on
both vertical and horizontal branches and may use both the same
crossing points between trees and the same routes within trees as
L.catta. In the reserve if a troop of either species moved quickly, it
tended to take the most "obvious" routes along large, unencumbered
branches, whether vertical or horizontal. In feeding both *P.verreauxi*
and *L.catta* ate fruit and leaves from the twigs of trees; both, therefore, ventured onto the smaller branches which grow in all directions. The difference of branch direction, then, is largely a question
of the sort of branch where the animals rest, which in *Propithecus* is
a vertical trunk or, more usually, a crotch with at least one vertical
branch to hold and a horizontal one to sit on. These observations are
true of the Mandrary gallery forest, where many of the unobstructed
routes lie along huge horizontal kily branches. A main kily trunk on
the other hand is so large that any lemuroid must dig its fingers into
the rough bark rather than clasp its hands around.

These generalizations also seem to be true of *P.v.verreauxi* in
the woods of Lambomakandro. My general impression of the wooded
region at Ankarafantsika, where Petter studied *P.v.coquereli,* is that
there are relatively more small trees than by the Mandrary. In a wood
with more saplings and fewer forest giants, *Propithecus* is more
likely to move on trunks than cross-branches. Similarly, in the wet
montane forest at Perinet, the larger trees grow with tall slim trunks,
as in most tropical forests, and *P.diadema* frequently leaps from
trunk to trunk. When a very large tree offers horizontal branches as
pathways, *P.diadema,* like *P.verreauxi,* hops along them. Finally, in
the extraordinary euphorbia forest the *Didiera's* vertical spires are the
only large trunks available, and *P.v.verreauxi* hops between them.
Therefore, *Propithecus* can and does exploit whatever branch type is
available, although it frequently moves and rests on vertical
branches.

BRANCH HEIGHT

As Petter (1962*a*) says, *Propithecus* and *Lemur* keep to much the
same heights when in the trees, although *L.catta* descends more
frequently to the ground. Tables II–2 and III–1 show the total time
spent at different heights by *P.v.verreauxi* and *L.catta* in the reserve.
There were, of course, seasonal and daily changes. The heights for
P.v.verreauxi correspond to the density of foliage, which was the
same as the density of their food. The bulk of the foliage in both

kilys and other large trees lay between 13 and 21 m with only the tallest kilys and *Celtis bifida* protruding to 30 m. Small trees and the lower branches of the larger ones had fairly dense foliage from 7 to 13 m and sparse leaves from 3 to 7 m. Below this lay only stems of saplings, bushes, and the ground.

P.*verreauxi* spent most of its time sleeping and feeding in the trees above 13 m. It came down to sleep in small, dense trees where it was well shaded—from 3 to 13 m during the day, especially in the hot season. It also came down to feed on particular trees in bud or in fruit. The difference in height between March and April of 1963 and of 1964 largely reflects the abundance of *C. bifida* berries, which usually grow between 7 and 13 m.

P.*verreauxi* occasionally fed on the tops or edges of bushes between 1 and 3 m, but at that height they mostly hopped rapidly among the bare, unencumbered stems of saplings.

The animals descended to the ground only when they were unaware of, or completely indifferent to, observation. In 1964, following a hurricane, P.*verreauxi* came down to feed on fallen kily beans, one animal at least spending 50 minutes on the ground. The animals sometimes sat on the ground in the middle of the day feeding listlessly. Very tame *Propithecus* occasionally crossed large open spaces when they could equally well take an overhead route through the trees. Sometimes an animal bounced briefly on the ground in locomotor play. Finally, a troop in territory already occupied by another troop might take to the ground as an alternative route. This contrasts to the *L.catta,* which walked and fed on the ground nearly everyday.

It seems likely that the same remarks apply to branch height in other similar forests. In the Euphorbia forest, where dense vegetation only extends up to about 5 m, *Propithecus* must have a much smaller vertical range. In particular, it must seek shade at midday either in the branching center of a *Didiera,* about 3 or 5 m up, or on or very near the ground. The available food resources are so small that *Propithecus* must range from the ground to the *Didiera* tops in its search. I have, in fact, seen a *Propithecus* at the top of a 13-m *Didiera,* eating the nosegay of flowers which grew there, but no animals were tame enough to stay near the ground when I approached.

WATER

P.*v.verreauxi* probably do not need to drink water. I have not even seen them lick leaves after a rain, nor have I heard other people say they drink. Their survival in the euphorbia forest without access to gallery forest indicates they can tolerate extreme drought.

FOOD

Table II–3 lists *Propithecus'* food plants in the reserve and Table II–4 those at Lambomakandro. In the reserve the kily (*T.indica*) is the staple throughout the year. There was a high proportion of *C.bifida* berries in March and April and of *Acacia* buds in August of 1963; but in March and April of 1964, when there had been less rain than in 1963 and a hurricane shook down the kily beans, *Propithecus* ate less *Celtis* berries and kily and more leaves of other trees and *Albizzia* seeds.

Feces of *Propithecus* largely contain undigested kily seeds. It seems that, as with *P.v.coquereli* at Ankarafantsika (Petter 1962*a*), most of the animals' nourishment comes from the thin seed covering rather than the large beanlike seed itself.

P.verreauxi ate the fruit of about 55 per cent of the species of its food plants during about 65 per cent of the observations (Table II–5). It ate the leaves of 70 per cent of the species but during only 25 per cent of the observations. It ate few buds or flowers, except in August, when *Acacia* buds became a staple. Thus its food habits resembled those of *L.catta,* in spite of its specialized leaf-eater's digestive system (Hill 1953).

Petter (1962*a*, 1965), Boggess and Smith (Buettner-Janusch, 1966), and I have never seen *Propithecus* eat insects or other animal matter. They will not accept meat or insects in captivity (Hill 1953, Andrew, personal communication).

At the same season of the year, during March and April, 1963, *P.v.verreauxi* at Lambomakandro were observed to eat twelve species of plants (Boggess and Smith, personal communication) and in the reserve were observed to eat eighteen species. Of these, only one species, *Albizzia bernieri,* was the same. (*Canthium* sp. was eaten at Berenty the following year.) However, *P.v.coquereli* in August at Ankarafantsika ate kily and *Ficus* spp., like *P.v.verreauxi* at the other end of the island (Petter 1962*a*). It seems that the various subspecies are not differentiated by specific food requirements but eat whatever the forest has to offer.

SUMMARY

Propithecus, like other primates, seems to be an opportunist. It eats leaves, fruit, or flowers and can feed on a different range of vegetable species in different forests. It can move on any type of branch, at any

TABLE II-3

Propithecus verreauxi; Food Plants Eaten in the Reserve

Plant	Family	Fruit, Leaves, Inflorescence	Per Cent Observations			
			Mar.–Apr. 1963	June–Sept. 1963	Mar.–Apr. 1964	Avg
Acacia sp.	Mimosaceae	LI	–	20	14	11
Albizzia bernieri	Mimosaceae	FL	1	–	14	5
Canthium sp.	Rubiaceae	F	–	–	3	1
Celtis bifida, C.gomphophylla	Ulmaceae	F	38	–	20	19
Crataeva greveana	Capparidaceae	L	1	1	–	1
Dioscorea sp.	Dioscoreaceae	L	5	–	–	2
Ficus cocculifolia	Moraceae	F	1	–	–	0
Ficus grevei	Moraceae	FL	1	–	3	1
Ficus tiliefolia	Moraceae	FL	5	–	6	4
Mazunta modesta	Apocynaceae	FL	5	2	–	2
Melia azedarach	Meliaceae	L	1	–	–	0
Melothria sp.	Cucurbitaceae	L	1	–	–	0
Phyllanthus sp.	Euphorbiaceae	L	–	1	3	1
Pithecellobium dulce	Mimosaceae	L	1	–	–	0
Poivrea sp.	Combritaceae	FL	–	3	–	1
Rinorea greveana	Violaceae	LI	5	2	6	4
Scutia myrtina	Rhamnaceae	L	3	–	–	1
Solanum sp.	Solanaceae	FLI	2	1	11	5
Tamarindus indica	Caesalpiniaceae	FLI	23	66	14	34
Tarenna grevei	Rubiaceae	FL	1	1	–	1
Tricalysia sp.	Rubiaceae	LI	5	–	6	4
?		F	1	1	–	1
?		L	–	1	–	0
?		L	–	1	–	0
Total observations			100	100	100	98

TABLE II–4

Propithecus verreauxi; FOOD PLANTS IN LAMBOMAKANDRO,
Mar.–Apr., 1963

Plant	Family	Fruit, Leaves, Twigs
Albizzia sp.............	Mimosaceae	F
Bathiorhamnis louveli	Rhamnaceae	LT
Canthium sp.	Rubiaceae	L
Execercaria sp..........	Euphorbiaceae	L
Grevia barorum..........	Tiliaceae	L
Grevia saligna...........	Tiliaceae	L
Salvadoropsis sp........	Celastraceae	L
?................	Acanthaceae	L
?................	Euphorbiaceae	L
?................	Euphorbiaceae	L
?................	—	L
?................	—	F

TABLE II–5

Propithecus verreauxi; FOOD PLANTS SUMMARY

	Mar.–Apr., 1963	June–Sept., 1963	Mar.–Apr., 1964	Total
(1) TOTAL SPECIES FOOD PLANTS........	18	12	11	24
(2) TOTAL OBSERVATIONS OF FEEDING....	84	134	35	253 Avg
(3) PART OF FOOD PLANT EATEN (%)[1]				
spp. leaves eaten.................	85	90	80	85
spp. fruit eaten..................	55	50	65	55
spp. flowers eaten	20	25	45	30
obs. feeding leaves...............	20	10	45	25
obs. feeding fruit................	75	70	50	65
obs. feeding flowers..............	5	20	5	10
(4) *P. verreauxi* FOOD PLANTS ALSO EATEN BY *L.catta*				
spp. eaten by both *P.verreauxi* and *L.catta* as per cent of (1)..........	70	60	85	70
obs. *P.verreauxi* eating these spp. as per cent of (2)	90	95	95	95

[1] Percentages do not add to 100 because animals often ate more than one part of species of food plant.

height, and occasionally descends to the ground. There is nothing to indicate that *Propithecus* or its subspecies are confined ecologically to any particular type of forest, although *P.v.verreauxi* may be the only form which can tolerate the euphorbia desert.

The major limitation is that *Propithecus*, like other lemurs, is confined to forested areas and cannot live on open ground. A second limitation is that *Propithecus*, unlike the small nocturnal lemurs and unlike such higher primates as chimpanzees (Goodall 1963, 1965) and baboons (DeVore and Hall 1965; Altmann, personal communication; T. Rowell, personal communication), apparently does not eat meat or insects.

The most interesting aspect of *Propithecus'* ecology is that troops defend small territories of nearly constant boundaries and fairly equal size. A *Propithecus* group thus neatly marks out for the ecologist the total area of forest which it is using and a minimum area which it shares with no other group. Petter (1962a) states that in disturbed areas of forest, *Propithecus* make long detours around clearings to reach trees they were in the habit of frequenting. I have seen animals sitting among fallen *Didiera* trunks in recently cleared land. Both of these observations confirm the *Propithecus'* attachment to their familiar range.

Propithecus are notoriously difficult to keep alive in captivity. As suggested by Webb (Hill 1953) the remedy may be to give the animals greater freedom and a wider variety of diet rather than to feed them any specific food.

INDIVIDUAL BEHAVIOR

GENERAL AND DAILY ACTIVITY

P.verreauxi have a low level of general activity. Although they move in long ballet leaps, they spend most of the day sitting.

As mentioned above, 100 m is a major displacement, and a troop would rarely travel more than 200 or 300 m in a day. Individual animals move little more than this, except for inevitable hops from one branch to another for food.

During a typical day, a troop will move a little in the sleeping trees at dawn and progress 20 to 50 m to sunning trees; at about 8 or 9 A.M. it will move another 20 or 50 m to feed. This is repeated about noon before the siesta. Between 2 and 4 P.M. the animals progress to the evening feeding spot, and after a longer progression, settle to sleep between 5 and 7 P.M.

INTELLIGENCE AND SENSES

Propithecus does not manipulate objects in the wild and does not carry or pick up anything but food or food-bearing branches. It may hop a few feet with a fruit in one hand before sitting down to eat the fruit. In general, prosimians have a very poor understanding of unfamiliar objects (Jolly, 1964*a, b*); and *Propithecus,* from its behavior in the wild, seems to resemble *Lemur* more than monkeys in this respect.

Propithecus' sense of vision may be acute. They never failed to see me if they looked in my direction, even when I was still and partially concealed. Furthermore, variation among their own faces is so marked that they may well recognize individuals on sight. They hold their muzzles downward, staring forward with both eyes; so vision must be stereoscopic. The vivid coloring of different races strongly leads to the presumption that they can see in color, at least at the orange-maroon end of the spectrum. Their sense of smell must also be acute, for they leave their individual scents on branches, and members of other troops will come and sniff at the branches. I could not smell these marks even at 1 m away.

POSTURES AND LOCOMOTION

Propithecus usually keeps its body vertical, both in locomotion and at rest. It hops with both hind legs, which can propel it into a soaring 10-m leap. On large horizontal branches or on the ground it bounds along in a series of hops, with body upright and arms held upward like a walking gibbon's. In fast hopping, the arms flail in circles, moving down with each spring. *Propithecus* will, however, frequently walk quadrupedally along the top of a branch; less often it descends a nearly horizontal branch by walking quadrupedally underneath it. On occasion *P.verreauxi* walks bipedally, placing one foot in front of the other. It may even brachiate, swinging under a branch for three or four handholds.

This vertical posture is, of course, reflected in the animals' anatomy. The huge jumping thighs are the most obvious characteristics. The feet are also very large and powerful with strong hallices that can grip a branch to hold *Propithecus'* weight while it is feeding, and, even more important, catch and hold the branch where the animal lands at the end of a leap. The arms are relatively small and the hands long and hooklike since they usually reach up so the animal's weight is hanging from them rather than supporting him from beneath, as in quadrupeds.

It is interesting to compare *Propithecus'* anatomy, particularly the hands and feet, to that of brachiators such as the gibbon. There are similarities which arise simply from the direction of the weight, although the gibbon's propulsive force is its arms, while in *Propithecus* it is the hind legs.

SLEEPING

P.verreauxi sleep in branches or in crotches of trees, usually about 13 m from the ground. The animals may sleep singly or with one or more of their troop in close contact. If alone, they sit on a fairly horizontal branch and hold with hands or feet to a more vertical one. The vertical one is usually a handhold, not a backrest. If several sleep together, one animal will hold a vertical branch, while the others grasp the first animal.

In general, any large tree seems to be suitable for sleeping as long as it is within the troop's undisputed territory. The troop may return to one tree for three or four successive nights, then shift erratically to a different tree in another part of the territory. There are no particularly favored or comfortable sleeping branches; even when the troop returns to the previous night's tree, the animals choose different crotches to rest in.

P.verreauxi wake about dawn and usually move a little. The group breaks up and re-forms on another branch. Animals may feed briefly. They may then drop back to sleep until as late as 8 or 9 A.M. They may, on the other hand, begin to sun themselves almost with the first light from 6:00 to 6:30 A.M. Early morning movements depend, though erratically, on the weather. On a very wet morning *P.verreauxi* generally stay huddled together or hunched over singly until quite late, but they generally attempt to sun themselves in any pallid rays which break through. If it is hot, *P.verreauxi* wake early, sun briefly, and start to feed. In the cold season, from June through August, a river mist fills the gallery forest until 8:30 or 9:00 A.M. *P.verreauxi* sleep late, then begin to feed a little, and settle again to sun for an hour as soon as the mist clears. When the troop moves from its sleeping branches to its sunning branches, the animals first urinate and defecate; then the troop leaps to the sunning branches in close formation.

SUNNING

There are no specific sunning places, but the sunning place on any one morning is less than 50 m from the previous night's sleeping place. There is less choice of sunning than sleeping spots, since they must be large branches where the animals can lean back, often without hanging on, fully exposed to the eastern light. Almost any crotch will serve for sleeping. Usually, the sunning spot is a dead tree or an opening in the foliage of a large tree where sunlight can reach sizable branches. A few sunning spots are as low as 7 m above ground, but many lie above 21 m.

The troop arranges itself on the sunlit branches. Two animals are occasionally in contact, but more often each one is separate. They spread their legs and hold their arms horizontally to their sides; so that the sun warms their sparsely furred belly, underarms, and thighs. Their heads loll to one side, with eyes closed and squinted against the sun. When a *Propithecus* is thoroughly toasted on one side, it slowly turns around and spreads again until its back is also warm. Only rarely does an animal eat while sunning.

Many Malagasy tribes tell legends of this sunning. They say that the lemurs worship the sun and even that they are incarnations of the ancestors who continue their worship in this animal guise. It is difficult to watch a sunning lemur without being anthropomorphic, but to Western eyes it seems less like religious fervor than like our indolent cult of Sunday at the beach.

SELF-GROOMING

As the *Propithecus* warm up, they begin to groom themselves, scraping their fur with the lower teeth. The lower canines and incisors of all Lemuroidea (and Lorisoidea) are prominent and narrow, like the teeth of a comb. In fact, lemuroids use the structure more as a scraper than as a comb, alternately licking and scraping their fur with upward movements of the whole head (Buettner-Janusch and Andrew 1962). The repeated upward head movement is unmistakable from any angle. *Propithecus* do not use their hands to pick up fine particles; so their grooming does not have the same gesture as that of monkeys (Petter 1962a). They hold on to the area being groomed with their hands but do not part the fur. Adult *Propithecus* do not clinch their fingers around tufts of fur, a gesture which Bishop (1962) believed might be a pre-adaptation for fine control of the hand. Instead, their long curved hands can only clasp around their limbs to assist in grooming.

The *Propithecus* may drop their heads on their breasts to groom their stomachs, arms, or legs or they may twist around to groom the small of their backs or pick up their spindly tail in both hands to groom its end. Very frequently they groom their genitalia, licking and worrying with the tooth-scraper.

Self-grooming generally finishes with a good scratch with the toilet claw. The second toe of each hind foot bears not a nail but a claw which sticks up at right angles to the toe. Lemuroids scratch themselves with this toe just like a dog, on the shoulder or behind one ear. They use this claw also to clean their ears, by carefully inserting the toe and turning it back and forth. Then they remove it from their ear, hold the foot briefly in front of their eyes and nose, and lick the claw.

During this period of sunning and grooming, *P.verreauxi* may also groom each other or wrestle and play (see pp. 76–79).

URINATION AND DEFECATION

When the troop is ready to move again, the animals urinate and defecate. One begins and the others follow suit before the first has finished. This simultaneous urination and defecation by several troop members is frequent enough to be more than coincidence.

When a *P.verreauxi* urinates or defecates, it lifts the base of its tail and holds the outer part roughly horizontal, in a position used only during excretion. Its anus is behind the horizontal branch where the animal stands or well away from the vertical branch, so that feces fall to the ground without fouling the branch. Females may urinate in this position, but males generally urine-mark a branch instead.

As with other primates, lemuroids have no particular sites where they excrete but do so from any branch where they are standing. As with other primates, any emotion or fear provokes a barrage of urination and defecation. After the animals have been resting, feeding, or sunning in one spot and before they prepare for locomotion to another spot, they also urinate and defecate.

They urinate and defecate if mildly alarmed by an observer, after moving overhead to see better or "sifaka" (see pp. 58–60). The joint effect may be unpleasant for the observer, but there is no reason to think this result premeditated as with some New World monkeys (Carpenter 1934).

FEEDING

At some time between 7 A.M. (in the hot season) and 10 A.M. (in the cold season), the *P.verreauxi* move from sunning to feeding trees. The group feeds seriously for 2 or 3 hours. Usually an animal will eat for 10 or 15 minutes, then rest, then eat again; but occasionally it will eat without stopping for an hour or more.

The feeding trees may be located in any part of the territory. They may be near the sunning place or as much as 100 m away. The group usually feeds in one large tree at a time but may cover two or three trees. Generally they move from one kind of tree to another or to a low sapling or liana during the feeding period. Not all individuals move at once; so the group is widely spaced. Most territorial disputes take place during the morning and afternoon feeding periods, since the groups then are moving and likely to meet each other.

P.verreauxi feeding postures are as varied as its locomotion. In large-branched trees with dense foliage, it sits upright on a horizontal branch or clings to a vertical one and reaches out with one hand for a twig, which it draws in to pick off the leaves or fruit with its muzzle. On small branches, though, *Propithecus* often slings itself under the branch like a sloth; the branch may bend almost vertical with the animal's weight as it picks off the outermost leaves. *Propithecus* can feed for 5 or 10 minutes at a time while hanging vertically from its feet alone.

In kily trees, the *P.verreauxi* can sit on large branches and still reach the dangling seed pods. The animal reaches out for a pod-bearing twig and hooks the twig toward itself. It holds the twig so that the pod is to its nose and smells up and down the pod. In May, when most of the pods are old and eaten out by insects, *Propithecus* may reject five or six before biting into one. It takes the pod in its molars and bites it off. It never breaks the pod loose with its hand, unless accidentally while drawing in the twig. It often, however, holds one end of the detached pod in a hand while chewing the other end like a stick of licorice. The fleshy outer part of the pod is quickly chewed off and the first seed swallowed while the fragments of pod drop down. *Propithecus* seems adept at removing kily seeds to swallow them. The animals are less skillful at holding the chewed pod— usually about half of it drops to the ground. One of the easiest ways to find a *Propithecus* group, unmoving and invisible in a treetop, is by the sound of dropping food. A few kily leaves as well as seeds are eaten, but usually only when the pods are old and rotten, or as a brief mouthful of salad after several minutes eating seeds.

Other buds and fruits are usually smelled, then taken directly with the mouth. Most are too small to hold in the hand—only various Ficus fruits can be held and munched as we hold an apple. For the other fruits, *Propithecus* hangs in slothlike position or spread-eagled among fine twigs and must contrive to bring mouth to twig or twig to mouth. Lemurs in general are clumsy at holding objects (Bishop 1964). *Propithecus,* with its large hand, is exceptionally awkward and does not try to hold small buds and berries.

Most food is bitten and chewed with the molars, but fine buds may be picked with lips or tooth-scraper. *Albizzia* seeds are individually chiseled from their flat pods with the tooth-scraper.

EVENING SIESTA

At about 10 A.M. in the hot season, *P.verreauxi* descends to sleep in bushy foliage about 2 to 7 m above ground. It remains in the shade and practically disappears until about 4 P.M. The maximum temperature in January reaches about 43° C (110° F), and birds and people also tend to take siestas.

In the cold season, the *Propithecus* sometimes do not sleep at all during the day but remain, with reduced activity, 13 to 21 m up in the kilys during the early afternoon. They do move about noon from their feeding trees to another tree of the same height. Many troops, however, descend to smaller trees between 12:30 and 2:30, when they begin to feed again. In the cold season the animals may rest together, in groups of two or more, on open branches; in the hot season they remain singly, each in his own clump of leaves.

Siesta positions are about the same as night sleeping positions, although the animals occasionally sprawl flat on a branch with arms and legs dangling.

A second period of feeding and grooming, similar but shorter than the morning one, comes in the afternoon. *Propithecus* then moves from feeding to sleeping tree, often 100 m away. The troop is always settled for sleep by dusk, that is, 6:30 P.M. in the hot season. In the cold season, the *P.verreauxi* troop may settle as early as 4:30 to 5:00, although it may make occasional rearrangements on the same few branches. Once a troop, which had been feeding through the noon hour, settled at 2:30 P.M. and did not move until I left after moonrise, at 9:00 P.M. *Propithecus* rarely moves during the night, for the troop members are usually in the same crotches next morning.

RELATIONS WITH OTHER SPECIES

HAWKS

P.v.verreauxi has one group of aerial predators, the hawks; one group of annoyances on the ground, the humans; and a large number of species with which it shares the treetops.

Hawks and other raptors prey on *Propithecus* and are probably their chief natural enemies. I never saw a hawk actually attack a *Propithecus* troop, but the troops roar their alarm bark whenever they see a flying hawk. The call can be given for another object, such as a human observer, but only in very high-intensity alarm. Once, when I had frightened a *Propithecus* troop and a teasing *L.catta* blocked their escape, the leading *Propithecus* began a roaring bark at the *L.catta*. The usual stimulus, though, is a hawk (or an airplane). A *Gymnogenys radiata,* the harrier hawk, a magnificent bird of pale gray plumage, black-and-white speckled breast, and in breeding season coral-red feet and skin around the eye, caused the longest disturbance I heard. The bird first dug with alternate feet in a hole high up in a branch for 30 minutes. It managed to extricate some small creature which it ate—perhaps a lizard or fledgling bird, perhaps even a *Lepilemur* (Rand 1936). The nearest troop of *Propithecus* "growled," a low series of clicks made with closed mouths, softly throughout. Then, when the hawk flew, they erupted into roaring barks. I followed the bird's zigzag course for 2 more hours through the forest by the repeated roars of three *Propithecus* troops over which it crossed and crisscrossed.

More usually, the troops roared briefly at the silhouette of a high-flying hawk, or even one nearby. On one occasion a brown hawk landed between troops of *Propithecus* and *L.catta,* one tree from either. The *Propithecus* roared, the *catta* shrieked. Then while the hawk perched, both groups again fed calmly, only to shriek and roar the moment the hawk spread its wings to fly.

A *Propithecus* roars with its head back and mouth pursed into an "O" (orbicularis oris contraction). It inhales and exhales three or four times in a resonant bellow that carries at least two troops away. The roar may be of higher or lower intensity, that is, long or short; but it is always loud. It can be given with the head down, since I once saw a troop roar with noses down at a bird (not a hawk) that suddenly flew below them. This roar resembles the "alarm call," not the "song" of *Indri.*

One troop member may see the hawk first and begin the roar, but other troop members join at once, in perfect synchrony. They may look upward afterward in search of the trouble. Neighboring troops who hear the roar become alert and fidgety; they often look upward. If they are already alarmed, they may roar themselves; for instance, they may roar at the observer immediately after a troop in the distance has roared. If the hawk that alarmed the first troop flies over them, they are also likely to roar. A troop may roar two or three times within about 5 minutes but after that are likely to be silent, even if the same hawk keeps circling above.

I cannot guess how serious a predator the hawk is. Only one adult *Propithecus* I could identify disappeared from the area (probably died) between March and September; so there is surely no great predation on adults. However, a *Propithecus* infant of 2 or 3 months, exploring alone, would be at the mercy of a hawk.

HUMAN BEINGS AND GROUND PREDATORS

Human beings receive quite different treatment. *Propithecus* flee a human being only if he chases them, or if they have been hunted before. A naïve troop will stop when they see him and start to growl (low blurred click series) in their throats. If the human being stops too, they slowly begin to leap toward him. The boldest animal begins to "sifaka," the noise which gives the local name to *P.v.verreauxi*. The *P.verreauxi* stares straight at the intruder, eyes wide open, and gives a bubbling groan, with its mouth shut, ended by an abrupt syllable click with mouth open, lips covering teeth (Andrew 1963a, "chicken-like cluck"; Petter 1962a, "omb-tsit"). The *Propithecus* can chew during its first syllable. At high intensities the first syllable is prolonged and musical, the second loud; at low intensities either syllable may be dropped.

P.v.coquereli makes a similar "sifaka" that can be distinguished by ear from that of *P.v.verreauxi*. *P.d.diadema* gives a one-syllable call, which I think combines both elements of the sifaka, and certainly ends with the loud click.

In eleven cases a male was first and boldest of the "sifakers," in six cases a female, and in two cases a juvenile. The rest of the troop follows suit, however, each animal sifaking in its own rhythm. Animals may sifaka sympathetically without seeing the cause of disturbance. Between sifakas they "growl" or "snore."

They also give head jerks. An animal flings its head backward sharply, so that its nose points upward or even back. This is a highly stereotyped gesture. It may be repeated three or four times, but between head jerks the *Propithecus* stares at the intruder. The gesture is possibly derived from an intention movement of jumping.

The troop members hop closer, in full view on low vertical stems. They come as close as 3 to 5 m, but they remain in a group, all facing the observer, each one usually on a separate tree trunk. They may repeatedly lick their noses and usually weave their heads from side to side while staring. As the group becomes more excited the sifakas become more and more prolonged, the final click more and more explosive. At this point many local legends say the *Propithecus* attack and eat human beings.

In reality if the observer returns the *Propithecus'* stare, they lose heart. An ordinary bout of sifaking lasts 5 to 45 minutes, although Petter (1962a) reports hearing an animal sifaka for 2 hours. The troop eventually retreats, hopping backward, still facing the observer. They continue to sifaka until they are out of sight.

Sifaking is presumably a way of mobbing ground predators. The native carnivores in Madagascar are all viverrids. One, the catlike fosa (*Cryptoprocta ferox*) grows to be a meter long and could certainly kill a *Propithecus* if it were able to catch one. The fosa originally lived throughout the island and so would have menaced all of the subspecies of *Propithecus*.

I only heard sifaking directed at people and given at very low intensity in the height of territorial disputes. I considered the possibility that it was largely a territorial vocalization given toward people as round-eyed, round-faced intruders on the territory. This can only be checked by studying *Propithecus* during the breeding season, when all disputes must be more intense. However, it seems likely that the sifaka is an ordinary mobbing call. This is supported by the fact that *L.catta* will spring up from the ground if nearby *Propithecus* sifaka. *L.catta* also have a well-developed, but quite different, mobbing call for ground predators, which was not used in encounters between different troops.

A *Propithecus* group sifakas only the first two or three times they are watched, or if revisited after a lapse of several months, and each time with decreasing vigor. Then they stop sifaking, and become scary and nervous, fleeing from observation. In the Mandrary region they grew tame enough to watch easily after only a month or two; elsewhere, of course, it would take much longer. Even in the reserve, they were never tame enough to touch since I did not feed them. They treated me like a member of another species or as they treated the *L.catta*. They fed, played, and groomed while I watched, but they glanced at me frequently and moved farther off if I approached or brandished a camera. Most social interactions took place at least 10 m away from me, although they often fed closer.

BIRDS AND BATS

Many birds, the flying fox (*Pteropus rufus*), and *L.catta* share the treetops with *P.v.verreauxi*. Several birds, of which the gray parrot *Coracopsis vasa* is most obvious, probably eat the same foods. Parrots crack kily nuts with their beaks and may feed simultaneously with a troop of *Propithecus*. *Propithecus* never mistook parrot for hawk, although they occasionally roared at other birds.

Usually *Propithecus* ignored birds other than hawks, but my general impression is that bird parties did not feed in trees already occupied by lemurs. One old male *Propithecus* was feeding near a trio of lovebirds (*Agapornis cana*). A crow (*Corvus albus*) burst from the foliage just over his head; the *Propithecus* started, swung under the branch where he was feeding to hang by all fours, and roared at the crow. Then he clambered up on the branch to begin feeding again but kept breaking off to peer at the tiny fluttering lovebirds.

A few bird calls caught the *Propithecus'* attention; twice the rattling call of *Cua gigas* and *Cua cristata,* once the guinea hen's cackle of alarm. On this last occasion, the guinea hen could be heard approaching far down a path. The *Propithecus* I watched were sunning as a group of five animals in contact with each other. As the guinea hen passed a troop of *L.catta,* the *L.catta* began to yap. The *Propithecus* alerted at the first cackle and stared down the path as the *L.catta* began yapping. Finally the hen came into sight, driving before her a black mongrel dog about four times her own weight. The *Propithecus* group, 70 feet overhead, sprang apart like popcorn, each animal to a separate branch. They made no sound, however. The hen and her victim proceeded up the path; a second *L.catta* troop took up the hue and cry, and the five *Propithecus* resumed their furry embrace.

Eight trees in the forest held a huge roosting colony of *P.rufus,* the flying fox. The *Propithecus* simply avoided these trees, although they moved freely through the neighboring ones. The bats probably competed with lemurs for food, since I saw bats settle in a fruit tree (*Ficus tiliafolia*) 5 minutes after the *Propithecus* left it at dusk.

Lemur catta

The species with which *P.v.verreauxi* had most contact was, of course, *L.catta.* For 80 hours of a total of 677 hours' observation, I could see both *Propithecus* and *L.catta,* and they could see each other. They almost always ignored each other, moving on their separate paths. Sometimes they would feed in the same tree, as little as a meter apart, with no apparent interaction or only a glance at the other animals. I occasionally saw one so close to the dangling tail of the other that I waited to see a tail pull. Sometimes two troops would cross at the same narrow gap between trees, *Lemur* alternating with *Propithecus.* The only consistent reactions were that, when one species gave its hawk alarm, the roar or the scream, troops of the other species alerted and looked upward or even leaped lower in the trees. *Propithecus* also looked about sharply when *L.catta* began to bark (the *L.catta* mobbing call) nearby. When I approached unfamiliar troops feeding together, both would approach to mob me in unison— *L.catta* barking, *P.verreauxi* sifaking.

The ecological similarities between *P.verreauxi* and *L.catta* are discussed in chapter V.

I saw only eleven true interactions between individuals of the two genera; in eight *L.catta* took the lead.

On February 28, 1963, I had startled a troop of *P.verreauxi* and one of *L.catta* that were sharing a siesta together. Single file they crossed the path on a large branch, about 4 m from the ground, the animals following perhaps 1 to 3 m behind one another. One *Propithecus* stopped to stare at me; the *L.catta* behind nearly bumped into him. The *Propithecus* simply raised one hand to chest height—a gesture which in all lemurs means an intended cuff. The *Lemur* drew back a step; the *Propithecus* dropped its hand; the *Lemur* made a square jump over the *Propithecus'* head, which is normal *Lemur* fashion of passing another animal without touching it. In this interchange the animals of two different genera reacted to each other's slight intention movements, and so could treat each other like members of the same troop.

On March 21, 1963, four *L.catta* entered a dead tree where two male *P.verreauxi* were sunning. The *L.catta* came up to a branch below the *Propithecus;* one actually stretched out its muzzle as if to smell the *Propithecus'* anogenital region. Again the *Propithecus* lifted one hand and the *L.catta* retreated. The *Propithecus* backed down its branch and hopped off without glancing at the *Lemur.*

On June 11, 1963, a subadult male *L.catta* chased a feeding old male *Propithecus* for two leaps, with the *Lemur* giving the defensive "spat" call.

The first troop of *P.verreauxi* that I saw at Berenty, February 11, 1963, and the first that I know saw me, became alarmed, sifakaed, and growled at me. They moved off, but a group of *L.catta* barred their path. Four *L.catta* had been playing jump-on-and-wrestle on a springy branch. One of the *Propithecus* leaped on the branch, only to be faced by an adult *L.catta*, who leaned forward and cuffed the air in front of the *Propithecus.* The *Propithecus* promptly swung under the branch and dropped. A second *Propithecus* jumped to the far end of the branch, swinging it down in a long arc; then, on the upswing, it leaped to land directly in front of the *L.catta.* The *L.catta* feinted again with head and body. The *Propithecus* flinched, then suddenly lifted its head to give an alarm roar. The whole *Propithecus* troop roared with him, and the *L.catta* fled precipitously. The *Propithecus* troop crossed out of sight, after which the *L. catta* reoccupied their playground.

On August 16 and August 24, 1963, and, in more leisurely fashion, on March 23, 1964, a whole troop of *L.catta* barred the *Propithecus'* way, while the *Propithecus* returned their teasing. Again, the animals leaped toward each other, stared, feinted approach, but never came into contact. All the game lay in leaps and counterleaps, the *Propithecus* trying to pass through the *L.catta* troop, the *L.catta* attempting to keep in front of them, facing the other direction. Since there are about twenty *L.catta* to five *Propithecus*, the *L.catta* had an advantage: if one animal does not out-guess the *Propithecus'* next move, another can do so.

In many ways the game resembles a *Propithecus* territorial dispute, but in these one contending troop drives the other backward; here sides mingle and eventually cross. Also, in territorial disputes the animals move in long leaps in one direction; in the teasing games they bounced all over the trees.

On March 23, 1964, an *L.catta* troop passed through a *P.verreauxi* troop in much more decisive fashion. The *L.catta* had wandered along a path on the ground more than 100 m outside their normal range. They then climbed to 40 feet above ground and began very rapid progress back toward their range, plowing through the midst of some resident *Propithecus*. One old male *P.verreauxi* just moved aside 1 m and watched the

L.catta stream past, but another male twice jumped in front of the *L.catta*. However, his shoulders were slightly hunched, his neck retracted, his gesture of incipient cuff slightly too high and too fast for conviction, as though it might be instead an incipient withdrawal. The *L.catta* continued to bear down on him, and each time the *Propithecus* sprang aside at the last moment.

Occasionally the *Propithecus* reversed the roles. On April 3, 1964, a *Propithecus* gratuitously supplanted an *L.catta*, although there was no other interaction between the troops. On March 23, 1964, an *L.catta* troop was sauntering down a trail when a male *P.verreauxi* landed on the trail, hopped for seven hops through the *L.catta*, dropped into "panther" position on all fours, stared at the *L.catta*, then hopped slowly off. On February 22, a *L.catta* troop crossed below some feeding *Propithecus*. The adults continued to eat, but one juvenile, already larger than a *Lemur*, descended and cut off the last member of the *L.catta* troop. This was an adult male, heavy and powerful, although traveling in subordinate position in his troop. The *Propithecus* bounced back and forth in front of him, while the *Lemur* tried to join the others. Even when the *L.catta* managed to get past, the young *Propithecus* followed, on his tail, so that the *L.catta* had to keep stopping to threaten him. At last the male *L.catta* escaped to his own troop and the *Propithecus* returned to join his placid elders.

TROOP STRUCTURE AND INTERTROOP BEHAVIOR

TROOP STRUCTURE

In 1963 in the Berenty Reserve, I counted fifteen *P.verreauxi* troops, which averaged 5.0 adults and juveniles per troop. This includes two troops of eight animals each. In 1964, I counted ten "troops," including one solitary male and one troop of eight animals, which averaged 4.6 adults and juveniles. The two exceptional troops at Bevala had eight and ten animals in 1964. At Lambomakandro the troops ranged from two to six animals, the average of five troops being 4.2 animals (Boggess and Smith, personal communication). At Ankarafantsika, Petter (1962*a*) reports an average of 4.0 *P.v.coquereli* per group for twenty-seven groups and for seven well-studied groups an average of 4.3. The latter figure is probably the more accurate, since it is easy to miss one or more individuals in the foliage.

I knew the composition of only ten troops at Berenty, a total of forty-six animals, before the birth of infants in 1963 (Table II–6). Of these, twenty-two were males, fifteen females, and eight juveniles. Ten groups in the same area in 1964 had twenty-four males, seventeen females, and five juveniles.

The high number of males is surprising. Total numbers of each sex are not significantly different. However, in each troop the number of males is equal to or greater than the number of females. The probability is less than 2 per cent of finding no troop with more females, in ten troops of this size, if the true sex ratio were 1:1.

TABLE II-6

Propithecus verreauxi; COMPOSITION OF TROOPS

Troop	1963						1964			
	Male	Female	Juveniles	Subtotal	Infants	Total	Male	Female	Juveniles	Total
BERENTY										
BN.........	2	2	1	5	1	6	3	2	1	6
WP.........	3	1	1	5	0	5	3	2	0	5
DE.........	2	2	1	5	1	6	2	2	1	5
BP.........	2	2	1	5	1	6	3	2	1	6
SB.........	1	1	0	2	—	2	1	0	0	1
IR.........	1	1	1	3	0	3	1	1	0	2
DK.........	3	2	1	6	2	8	3	3	2	8
ST.........	3	1	1	5	1	6	3	2	0	5
RH (SE).....	3	2	1	6	1	7	1	1	—	2
PK.........							4	2	0	6
Total in ten Ha, ...	20	14	8	42	7	49	24	17	5	46
BE.........	3	1	1	5	?	5?				
Total for 10 troops...	23	15	9	47	—	—	24	17	5	46
BEVALA										
Troop 1 [1].........	3	4	0	7	2	9	2	4	2	8
Troop 2 [2].........	—	—	—	7?	1	8?	6	3	1	10
LAMBOMAKANDRO										
Troop 1 [3].........	3	2	1	6	—	—	—	—	—	—
Troop 2.........	2	1	1	4	—	—	—	—	—	—

[1] Seven white, one melanistic

[2] Four white, six melanistic

[3] Three white, three melanistic

First total is number in same area in two successive years. BE is BElial troop, outside main study area, not counted 1964. PK troop took over RH territory in 1964.

Therefore, there is a significant probability that any troop will have as many or more males than females (Jolly, in press).

The only other primate which seems to have more adult males than females is *L.m.macaco,* studied by Petter (1962*a*). In most higher primates the male/female ratio lies between 1 : 1 and 1 : 3. In those species in which there are more females than males, it is usually thought this comes from differential mortality of infants and juveniles. Petter (1962*a*) suggests that the inverted sex ratio in *L.macaco* arises the same way. *L.m.macaco* has strong sexual dimorphism: the males are black, the females golden. Petter, therefore, could see that among the juveniles there were more females, while among subadults (1 year olds) there was a great predominance of males.

Hall and DeVore (1965) show that in baboons, the sex ratio appears unequal because males mature later than females. This mechanism could explain the ratios of male subadults of *L.macaco,* but seems unlikely for the *Propithecus,* in which most 1963 infants were seen again as juveniles and juveniles as adults.

An alternative explanation, at least for *Propithecus,* is that males live longer than females. In 1963, seven of the twenty-two males were small, rather scrawny, with deeply or multiple nocked ears or scraped bare patches on their tails. Of these, five seemed to be in fairly good condition—active, aggressive toward me, and with dense coats of fur. One I did not see in 1964 because I did not find the troop. The other four were alive and vigorous in 1964. Two more males were balding on their elbows, knees, and thighs, and they had caps faded to a pale red-brown and deeply sunken faces. One of these animals disappeared, and probably died, in July, 1963. I could not find the other in 1964 and believe he may have died. A third ancient male, outside the troops counted, did survive until 1964. Of the fifteen females, none was apparently "old" like these males, although it might be more difficult to tell since the females do not have nocked ears, and therefore the only criterion is the animals' condition. It seems possible, though, that there are more males than females simply because of longer life span.

One infant was born in 1963 in each of five of the ten troops, two infants in one troop, and none in three troops; one troop is unrecorded. Of the two troops at Bevala, one had one infant, the other, two.

Petter (1962*a*) and Griveaud (personal communication) say that *P.v.coquereli* has only one infant in a troop each year, and I found this probable in eight troops of *P.v.coquereli* counted in the same forest in 1962, since six troops had one infant. The slightly smaller

size of *P.v.coquereli* troops at Ankarafantsika may mean there is only
one breeding female in a troop, while the crowded troops near the
Mandrary may remain in larger units, with more females and thus
the possibility of two infants in a season. Only seven of thirteen well-
known females bore young. This might mean females give birth
every other year or it might mean that some were still subadult, if
Propithecus, like *L.catta*, first breeds at 2½ years. This possibility
does not change the apparent troop structure or sex ratio significantly, as is clear from the 1964 troop count on the assumption that
the known 1963 juveniles were still nonbreeding.

Of seven infants born in 1963, I found five in 1964 as juveniles.
One I missed because I did not find the troop. The other probably
died, for its mother was still with her troop. Of the seven juveniles
whose troops I saw in both 1963 and 1964, five (two male, three
female) grew to adulthood with their same troops. It is possible that
the other two joined troops outside the study area or that they died.

SCENT-MARKING

The members of a troop extensively scent-mark their territory. Male
Propithecus have a long dark scent gland, whose secretion stains the
fur (Petter 1962*a*), down the ventral surface of their throats. They
place their chin against a vertical branch, then push their chin
upward in a convulsive jerk until the whole throat is pressed against
the branch. They remain a moment with neck concave and throat
against branch, then start over, repeating the gesture three or four
times. It is obvious to the eye when a *Propithecus* throat-marks; it is
also obvious to another *Propithecus'* nose. Other members of the
troop smell the spot and often throat-mark or urine-mark the same
place. Hours later, members of another troop may again smell and
mark the same branch.

Males urine-mark even more frequently than they throat-mark.
They spread their knees, lift the base of the tail, place the tip of the
penis against a vertical trunk, and waddle 50 cm up the trunk,
leaving a line of urine drops behind. Again, other animals may smell
and re-mark the place.

Very rarely—only three times during my observations—a male
may "tail wave" as it urine-marks. It wiggles the base of its tail in a
horizontal figure eight pattern, while the limp outer end flaps from
side to side in large, sinuous figure eights. The gesture is highly
characteristic and spreads the animal's scent in the air. This gesture
may bear some relation to *L.catta's* tail waving, which advertises a
male to other animals in the immediate neighborhood, that is, to

members of the same troop. If the gesture serves the same function in *P.verreauxi,* it is probably commoner in the breeding season.

Females also urinate against vertical trunks, but less often and in a less stereotyped fashion than the males.

Rarely, males mark with the anogenital region, like females. This starts with the same posture as urine-marking, but the animal presses its anogenital region to the branch and rubs up and down. Females clearly mark with the vulval area, but males seem to use the back of the scrotum or anus, about as in fecal-marking.

Fecal-marking is not very stereotyped. The animal simply passes the anus toward the branch and defecates. The process shades from true anal-marking to ordinary defecation, which does not foul the branch.

I have included scent-marking in the section on intertroop behavior, since scent certainly distinguishes a territory, even in the owner's absence, and a territorial dispute involves a frenzy of scent-marking, urination, and defecation. However, scent-marking occurs whenever the animals are excited—as when alarmed at me.

A male may smell the urine of a female and throat-mark it, working gradually upward until his nose almost touches her rump. She may then turn to groom him, or, more often, cuff him and spring off (Petter 1962c).

When a male smells a mark, or himself marks at high intensity, he may "grin" with half-open mouth, corners of mouth drawn back, lips covering or partly covering teeth. Andrew (1963a) suggests that the grin, which is a gesture of submissive greeting in both lemurs and higher primates, may be derived from a protective response, for example, the "rejection response" to strong odor. The grin certainly looks the same in both situations, but interpretation is difficult, for scent-marking in the wild is often "emotional" in dramatic social contexts, and the grin appears only with high-intensity marking or in a submissive social reaction (see chap. V).

Scent is certainly individual by analogy with other mammals such as dogs and people. The various males of a troop smell and re-mark each other's marks. Marking occurs whenever the animal is excited, as in territorial disputes, in sexual behavior, or probably in dominance behavior. It serves to identify and assert the individual whatever the context.

TERRITORIAL DISPUTES

Territorial "battles" are formal affairs composed of leaping, staring, and scent-marking with very low growls or in silence. Invading troop members stay in close formation, move hesitantly, smell and scent-mark the branches. If unopposed, they may spread out to feed but remain alert and on guard. If the resident troop appears, the invaders bunch together, face them, and scent-mark.

The residents hop very rapidly toward the invaders without stopping to scent-mark, unless they too are very near the edge of their territory. The invaders wait until the residents are almost upon them, then hop backward toward their own domain. The troops often seem to mix, with animals leaping about in chaos. Each group, however, keeps its own orientation, the troops facing their goals like sets of opposing chessmen. Rivals may even land back to back in the height of battle as each faces outward from his territory. At the climax the outraged residents may sifaka softly. A territorial battle may be very beautiful, since everything depends on a fast, formal pattern of movement, each animal occupying sections of tree rather than opposing individuals of the other troop. Each tense leap, there-fore, carries an attacker toward a particular undefended area of tree, *not* into contact with an enemy. The troops, therefore, move in reciprocal formations.

The invaders often retreat at once, but occasionally leap side-ways to another branch or tree before they are chased away. The only way to avoid defeat is redirection—two otherwise tame troops may sifaka violently at the observer for 15 or 20 minutes, then part in different directions.

Two examples illustrate variations of the basic pattern (see maps VI and VII).

In late March, the kily seeds were old and dry, and the *Propithecus* fed largely on unripe *Celtis* berries. The Widow Peak troop spent from 9:10 to 11:40 A.M. of March 19 gorging themselves in a small stand of *Celtis*. They then crossed to a kily north of the path. The Dent Ear troop appeared still further north, just beyond an area of low bushes. Mephisto, largest male of the Dent Ears, hung on an exposed liana, staring across at the Widow Peaks, who stared back. By 12:10 all the Dent Ears moved slowly north, and the Widow Peaks settled for a siesta.
Next day, March 20, the Widow Peaks again fed in their *Celtis* grove from about 9:30 A.M. and had a siesta there from 11:20 A.M. At 1:10 the Dent Ears appeared, working slowly southward. At 2:00 they finally jumped to a liana bank near the Widow Peaks and followed each other rapidly to a huge kily crotch. There all five Widow Peaks bunched about 1 m from one another, facing the Widow Peaks, while Mephisto and Dent Ear, a battle-scarred male, urine- and fecal-marked. The Widow Peaks growled very softly, then hopped rapidly toward the Dent Ears, which immediately retreated north. The Widow Peaks then stayed about 3 m behind the Dent Ears, but drove them 20 m out of the liana bank. The Widow Peaks then returned. All three Widow Peak males throat- and urine-marked, while the female marked with her genitalia.
On March 21, the Widow Peaks sunned from 7:30 to 7:40 A.M. on a

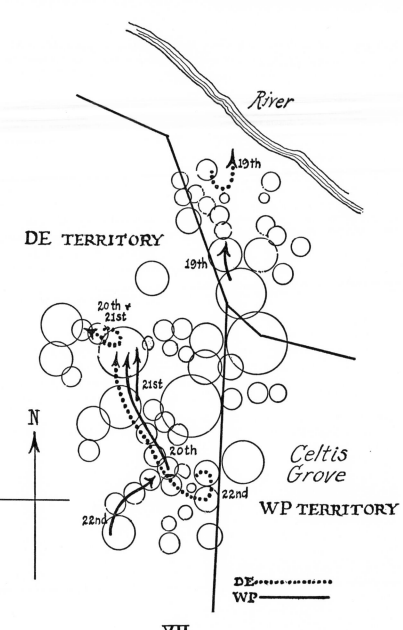

River

DE TERRITORY

19th

19th

20th
21st

21st

N

20th

Celtis
Grove

22nd

22nd

WP TERRITORY

DE••••••••••••••
WP━━━━━

VII

MOVEMENTS DURING "BATTLES"

MAP VII.—Confrontations of Dent Ear and Widow Peak *Propithecus* troops, March 19–22, 1963. Dent Ear range extended southward daily, finally including grove of foueting *Celtis* trees, although Widow Peaks chased Dent Ears northward in each formal "battle."

dead tree in the northern part of their range, then retired to the *Celtis*. Dent Ear himself promptly occupied the dead tree, while the rest of his troop sunned on the north side of its clearing. Between 8:30 and 8:35 they moved north, marking. The Widow Peaks started slowly north again and by 10:10 were feeding in the previously disputed kily. The Widow Peaks chased them out through the branches, then marked vigorously—but at 10:25 retreated south themselves, and at 11:00 crossed the trail eastward to feed there for the rest of the day.

On March 22, the Widow Peaks stayed in the south end of their territory in the early morning. But at 9:20 the Dent Ears appeared and began eating the Widow Peak *Celtis* berries, feeding in loose formation. At 10:20 the Widow Peaks surrounded by a group of *L.catta,* arrived from the southwest. The Dent Ears had a choice of three paths of retreat. They could either leap northwest, toward the attacking troop, or move east or south, still deeper in alien territory. They choose the first course, leaping fast and all together. The Widow Peaks pursued—but not too fast, not quickly enough to catch the invaders. The Widow Peaks delayed entering the northern kily until the Dent Ears safely left the other side. Then the Widow Peaks went into the kily for 10 minutes of marking and eventually moved southward.

Thus, during 3 days, the territorial boundary shifted south and the invaders established themselves in the *Celtis* trees while losing every battle.

Another hotly disputed area lay at the intersection of the Blaze Nose, Scrape Tail, and Red Head troops (map VI). Sideburn, a solitary male, lived here. Sideburn could keep no territory of his own intact, but moved from one spot to the next avoiding neighboring troops. Sometimes each troop occupied the corner of overlap in turn, one group moving away as another approached.

On July 14, the property rights were further complicated by the appearance of the Dent Ears to the south. The Blaze Noses sifakaed twice, and leaped eastward quickly, then leaped as quickly back and crossed to the north. A detachment of the Red Head group hopped westward, on the ground, the Dent Ears being above and behind; the Blaze Noses were above and in front. The Red Heads continued westward, below the 3-m level, to emerge beyond the battle. Meanwhile the Dent Ears arrived at 13 m, in full view, to land in a favorite Blaze Nose sleeping tree. The Blaze Noses circled, chasing the Dent Ears back again. Last to leave was Dent Ear himself, the aggressive old male. The Blaze Noses followed so fast that one actually landed on Dent Ear's back, grabbing with hands and feet. Then an extraordinary thing happened. The attacking Blaze Nose, as it clasped him, gave two upward scrapes with its head in the *grooming* gesture. They separated after no more than 5 seconds contact.

FORMATION OF TROOPS

Troops are relatively stable. Of thirty-five animals whose troops I saw in 1963 and 1964, I re-identified thirty-one. None of the thirty-one had changed from one troop to another. Two of the missing four were juveniles, the other two a male and a female that were either gone from their troops or changed beyond recognition. The period I was away included *Propithecus* breeding season; so it seems unlikely that troops are reorganized during the breeding season.

I saw only one troop formed, when a male and female of the Red Head troop joined the solitary Sideburn:

In March, Sideburn had a female of his own. They fed just north of the Blaze Noses, sandwiched between them and the Scrape Trails. Sideburn was one of the first *Propithecus* I learned to recognize, for his cap continued in a wide brown line down his cheeks, almost encircling his face. His female, on the other hand, was an undistinguished creature; so when she disappeared in April I could not tell if she died or perhaps joined the Scrape Tail or Door Knob troops.

Sideburn and the female weakly defended their patch of territory in March: on three occasions they throat-, urine-, and genital-marked, and stared at the approaching Blaze Noses for a long time before retreating. The Blaze Noses smelled and re-marked. Both groups once suspended operations to sifaka violently at me. After the female disappeared, Sideburn rarely or never throat-marked and rarely urine-marked. When another troop approached, he moved away. He had no territory of his own but was chivvied back and forth between Red Heads and Blaze Noses. This state of affairs continued until late June.

On June 26, Slitear, largest male of the Red Head group, was feeding with a female in a kily. The Red Head group itself only defended 0.2 hectares of large trees in their territory, as compared with 0.8 to 0.9 hectares for Blaze Noses, Dent Ears, and Widow Peaks, which may show the troop was already weak or unstable. Sideburn approached, closely followed, but not pursued by the old male, Redhead. Sideburn stopped when he saw Slitear ahead and then retreated, still followed by Redhead. Redhead, Slitear, and the rest of their group (one more old male, two females) moved west toward their own territory.

Next morning, June 27, Redhead and Sideburn sunned only 2 to 5 m from each other, in the acacias by trail 1. They both crossed the trail into a kily, where they fed.

On July 1, Redhead and Sideburn entered the acacias to sun, feed, and sit. The Blaze Noses had been in the same trees for the previous hour and were resting in small trees north of the trail. The Scrape Tails were also north of the trail to the east. Redhead throat-marked and Sideburn urine-marked sporadically; Sideburn sifakaed once. They stared in turn toward the Blaze Noses, the Scrape Tails, and me. The usual procedure for the neighborhood troops of *Propithecus* was to feed both in the acacias, and in a particularly luxuriant kily north of the path. Three crossing points 7 to 21 m above ground led from one to the other. On this occasion, though, the eastern route led within 8 m of the Scrape Tails, the western route through the midst of the Blaze Noses, and the direct route right over my head. I presented more of an obstacle than usual, standing up with movie camera and tripod. At 11:50 A.M., Redhead suddenly leaped onto the ground, crossed the path between me and the Blaze Noses in four hops on his hind legs, and entered the saplings and lianas on the north side. A minute later Sideburn followed by the same route. They leaped rapidly around on small trunks, less than 2 m from the ground, until they emerged in the kily where they fed.

Ten minutes later, though, Sideburn crossed to the nearest acacia, then stepped over again, cautiously to the nearest kily, started to feed, then leaped back with three jumps that carried him well into the acacias. There was a loud crash from the Blaze Nose area—something jumping or falling. Sideburn began to coo, the rare *Propithecus* contact noise. At

12:30 he inched back into the kily, 1 foot at a time. Someone gave low growls, but 2 minutes later Sideburn began to feed and thereafter fed in peace.

On July 6, Sideburn was ousted at 2:30 P.M. from the trees where he was resting by the Blaze Noses, who settled down there at 3:10 to sleep. Sideburn did not stop to contest the spot, but merely crashed away. He fed alone on July 8.

On July 12, a female from the Red Head troop sunned by herself at 4 P.M. to 4:30 P.M. above 70 feet in a kily overhanging trail 2. Sideburn joined her and was allowed to groom her very briefly. She put her arm around Sideburn's shoulders. Then she poked Sideburn's face with her nose. Sideburn "grinned," a gesture which routinely shows social inferiority in *Lemur* (Andrew 1962a) but which I only saw this once fully expressed by *Propithecus*. The corners of Sideburn's mouth drew straight back toward his ears, mouth slightly open, his teeth barely showing beneath the lips. Sideburn did not face her directly but remained in profile; then the two animals looked at each other a moment, still in contact. The female left, Sideburn looked after her and grinned again. Then he hunched over, chin on shanks between his knees, like a *L.m.rufus* I knew who was perpetually picked upon in a cage of four. Finally Sideburn groomed his stomach a little and began to feed in the kily.

Meanwhile, Slitear, the big male of the Red Heads, had been feeding alone with a female on July 8. July 12, these two, with Sideburn, crossed the trail on the ground to avoid the battle already described between the Blaze Noses and Dent Ears. They emerged to sun and feed southeast of the Blaze Noses. The female hopped to Sideburn, who first hunched over and then turned his face to her. She groomed his face. All three animals self-groomed. The female hopped off, while Slitear approached Sideburn, groomed, and left. Sideburn followed the other two.

After this, the troops stabilized themselves. Sideburn, Slitear, and the female were together on July 14, July 26, July 30, August 13, August 20, August 23, August 27, August 29, and September 2, although Slitear also was with the remaining Red Heads on July 26 and August 23. Redhead himself was alone on July 14 and July 29, after which I did not see him again.

The rest of the former Red Head troop, an old male, a female with infant, and a juvenile remained together and were seen July 14, July 26, and August 23.

In 1964, both old male Red Heads had disappeared. Sideburn was again alone; Slitear and his female formed a troop of two. The remaining Red Heads had drifted south out of the area. A strange troop, the Pea Keds, which first appeared in August 1963, took over most of the Red Heads' old territory.

It seems clear that new troops can be formed by a reassortment of adult animals from neighboring troops. It seems also clear that neighboring animals must know each other's sight and scent almost as well as their own troop members'. The males Redhead and Slitear took an active role in approaching Sideburn: presumably the males play as large a part in intergroup cohesion as they do in intragroup contact and grooming.

It also seems that when there is an inherently unstable situation in the troop structure, it may be solved, at least out of the breeding season, by amicable reassortment of neighboring animals. Thus, the geographical pattern of the neighborhood is maintained, and the high degree of inbreeding that must exist among animals of one region is also maintained.

SOCIAL BEHAVIOR

TROOP FEEDING AND LOCOMOTION

For most of the day, *P.verreauxi* remain a few meters from each other, without contact and with very few social signals (Tables II–7 and II–8). I saw only two hundred eleven intratroop interactions, or 0.8 per hour. (I was not there, however, during the breeding season.) This recalls *Colobus* or howlers, rather than *Cebus* or baboons.

The coo is a very soft sound consisting of muffled clicks against a tonal background and made with closed mouth. Animals give it rarely, when separated from the rest of the troop by a disturbing object such as the human observer or when there has been aggression in the troop. The coo is presumably a "contact" call and is derived from the infant discomfort call. I heard no long distance contact calls comparable to the wail of *L.catta* or of *Indri*.

There seems to be no social pattern in *Propithecus'* feeding. Any two animals may be near or farther apart. They do not resent another animal nearby. On three occasions I saw one animal take food from another's mouth by nosing toward it. Twice a juvenile took food from an adult male. Once the male made no reaction, but another male lifted its hand and caught the juvenile's hand. On the third occasion a female took food from an adult male. They again clasped hands, but whereas the male's movement toward the juvenile was abrupt, like an incipient cuff, the male and female seemed more coordinated, more like incipient play-wrestling. The troop may be spread as little as 3 m apart or as much as 15.

When the troop members move, they often follow each other in single file, although some members of the troop may take a different route. Again, there is no pattern of leader or follower, and any animal in the troop may initiate the movement.

FRIENDLY RELATIONS: CONTACT, GROOMING, PLAY (TABLE II–9)

Any two troop members may touch noses in "greeting," but the gesture occurs rarely. I saw animals groom each other only forty-nine times in the eight months before the birth of the infants, and I saw contact without grooming fifty-one times. Lemuroidea are very shy when grooming or in contact: it is easy to break up a clump of tame *Lemur* in the laboratory merely by staring at them. However, the

TABLE II-7

Propithecus verreauxi; VISUAL DISPLAYS

Sign	Accompanying Vocalization	Spp. with Homologous Signs	Occurrence
Stare...............	–, growl, sifaka	all Lemuroidea	to unusual objects, predators, other *P.verreauxi* troops
Lips pursed forward........	roar	*I.indri, L.catta*	to hawks
Mouth open slightly........	"fak" of sifaka	*P.diadema*	to ground predators
Mouth wide open, teeth covered...	–	*L.macaco, L.catta*	with play bite
Grin................	–	*L.macaco, L.catta*	to superior's approach, to threatened cuff
Tongue licking nose..........	–, growl	*L.macaco*	to observer (in uncertainty)
Head back..........	roar	–	to hawks
Head thrown back sharply.........	–, growl, sifaka	*P.diadema*	to ground predators
Head weaving	–, growl, sifaka	–	with stare
Duck head...........	–	–	to sudden bird movement or nearby roar
Groom..............	–	all Lemuroidea?	between any two animals
Hunch head and back.......	–, spat	*L.macaco*	to threat
Cuff, intention cuff........	–	*L.macaco, L.catta*	mild threat
Nose-poke...........	–	*L.macaco, L.catta*	to prevent animal from grooming infant
Touch noses...........	–	*L.macaco, L.catta*	rarely, in greeting
Nose rump...........	–	*L.macaco, L.catta*	Male smells urine, then anogenital area of animal higher on branch
Tail curled.........	–	–	at any time
Base of tail lifted........	–	–	while urinating or defecating
Tail lashed.........	–	–	rare, while urine-marking
Throat-mark.........	–	–	males, at any time, esp. in territorial disputes
Urine-mark...........	–	most Lemuroidea?	males or females, at any time, esp. in territorial disputes
Fecal-mark...........	–	–	males or females, any time
Genital-mark...........	–	most Lemuroidea?	females, at any time

TABLE II-8
Propithecus verreauxi; VOCALIZATIONS

Sign	French Equivalent [1]	Spp. with Homologous Sign	Occurrence
NOISES [2]			
low growl, snore........	ronflement	*L.macaco, L.catta* "clicks"	at other *P.verreauxi* troops, at distant or perching hawks, before or between sifakas
sifaka.................	omb-tsit, grognement	*P.diadema,* "vouiff," "vootch"	at ground predators
bark (roar)............	aboiement	*I.indri,* alarm call	at hawks, airplanes, and in great alarm at ground predators
CALLS			
infant squeaks..........	cri aigu	*L.macaco, L.catta*	infant groomed roughly or losing contact with mother
juvenile bubbly squeaks...		*L.catta*	juvenile disturbed by troop roar or sifaking
coo...................	roucoulement	*L.catta* twitter?	adult disturbed or separated from group, rarely while grooming
spat..................	cri de colere	*L.catta* spat	while cuffing or being cuffed

[1] Petter (1962a) for *P.v.coquereli,* whose vocalizations can be distinguished by ear from those of *P.v.verreauxi.*
[2] Rowell and Hinde (1962) define "noises," which have little or no tonal structure, and "calls," which have marked tonal structure.

normal frequency of contact, even when they are not observed, must be low in *Propithecus*.

Nevertheless, lemurs as a group are contact animals (Andrew 1964a, Bishop 1964). A lemur clump, when the animals sit with as much of their bodies in contact as possible, is the closest form of social communication. In *Lemur*, the result is a nearly spherical ball with three or four or five heads set into its surface and as many tails flung over the outside. *Propithecus* are too large to clump in every direction on a branch. Instead, they form "locomotives." One animal

TABLE II–9

Propithecus verreauxi; FRIENDLY INTERACTIONS IN TROOPS WITHOUT INFANTS

Pairs	Contact	Grooming	Play	Total Observed [1]	Total Expected [2]
Male-male..........	6	8	3	17	17
Male-female........	11	16	5	32	32
Female-female.......	1	–	–	1	4
Male-juvenile.......	7	6	–	13	20
Female-juvenile.....	15	7	–	22	12
Total identified..	40	37	8	85	85
Total not identi-fied..........	9	14	11	34	–
Total seen....	49	51	19	119	–

[1] Total friendly interactions per hour = 0.6.
[2] Expected on hypothesis that animals in *each* troop come in friendly contact randomly. Calculated here for all ten troops, but negligibly different if calculated for the four troops most often watched.
Low number of expected female:female encounters because there were normally fewer females than males in a troop—often only one female (Table II–6).

sits behind another, belly pressed against the back of the first and the big jumping legs enclosing the first on either side, very much like children playing train. As Petter (1962a) says the animals are "veritablement emboités." The locomotive may be only two animals in contact or grow to include the whole troop. It is, in fact, the position of maximum contact open to big, large-legged animals on a long branch.

Propithecus may sleep in contact, but a full-scale locomotive is more likely to form in the early morning, when the troop has just started to wake and move and before they spread out to sun. The only locomotive seen by Petter (1962a) was formed on a cold morning in August, but a locomotive may also form in the hot season or even at noon.

Propithecus, like other primates, groom each other (Plate 4). They usually sit facing each other to groom, although on occasion one will sit behind, above, or below or hang by its feet or adopt any other position physically possible.

Mutual grooming involves the same gestures as self-grooming: the upward scrape with lower teeth and licking with the tongue. Animals do not scratch each other with their specialized "toilet claw" on the second toe.

The groomee first presents the part of his body he wishes groomed, usually chin or underpart of upper arm. The groomer wraps his hand around this part, but does not clench his fingers in the fur. A favorite grip is one hand around the other animal's muzzle or even the flat of the palm on the end of the groomee's nose to lift the head for grooming. Then the groomee reciprocates so that roles are reversed every minute or 30 seconds.

The grooming may end quickly. If it continues, the animals alternately groom each other's genitalia, the head of one almost buried in the crotch of the other. This also is reciprocal. Each animal attempts to reach the other's genitalia, until the alternation breaks down and they are wrestling simultaneously for the privilege of grooming each other.

Propithecus adults wrestle and play. If the bout develops from grooming, the animals sit facing each other, sparring with both hands and feet. The object is usually genital grooming, but the animals grab wildly at each other's wrists and ankles, attempting to penetrate the other's guard. The play-wrestling does not necessarily develop from calmer grooming, but may begin at once.

Often they wrestle in other positions. Both animals may drop simultaneously to hang from a branch by their feet and spar with their hands, or to hang by their hands and kick. Wrestling is always mutual: both animals fumble simultaneously for a hold on each other's wrists, ankles, and noses. If an animal is approached to play when it does not wish to, it simply springs far away in one leap, and it is not pursued.

Propithecus play is irresistibly funny, but not because it is violent. The rough-and-tumble chases of young baboons look frenetically active compared with *Propithecus* play. The effect comes from the slow-motion, almost stylized fumbling for each other. *Propithecus* seems to savor every moment of the play, to go slowly enough so that each gesture will be enjoyed without achieving any real advantage over the adversary. Where baboons tease, dominate, defeat, play King of the Castle, *Propithecus* literally play the child's game of bicycling —when two children put their feet together and pedal around and around in perfectly balanced opposition.

Adult males frequently groom each other, sit in contact, and even play with each other. Of the eighty-five instances in which I could identify both animals (before grooming centered on the new-born infants,) seventeen were two males. This is precisely what we expect with random assortment, if animals groomed any troop member with equal frequency (Table II–9).

A male and female groomed, sat in contact, or played about twice as frequently as male with male. The reciprocal nature of all these activities meant that the male manipulated the female as actively as she did him. Only once did I see a female sit in contact with another, and this was as part of a locomotive that included the whole troop. I never saw two females groom each other. Of course, in only five of my ten troops were there two females, but I watched three of these troops intensively.

Similarly, of twenty instances of contact, grooming and play between adults seen by Boggess and Smith (personal communication) at Lambomakandro, ten were male-male, eight male-female, and three female-female.

In most primates, the females groom each other and take the lead in grooming males. *Propithecus* reverses the situation. Males groom each other and females; females groom males; and females rather ignore each other. The all-important social bonds of contact and grooming are equally forged by males and females.

DOMINANCE AND AGGRESSION

There is no dominance order in a *P.verreauxi* troop when the females are neither in estrus nor carrying infants. When the group moves, any member may lead. There are practically no aggressive interactions between animals (Table II–10). The usual sort of threat that I saw, for example, occurred when a male nosed beneath a female to smell her anogenital area, and she turned and lifted her hand as if to cuff him. Once halfway through the movement the cuff turned into a nose hold for grooming. The female then looked at me and hopped off.

I did, however, once see a female chase a juvenile, abetted by the rest of her troop. In fact one member of the troop repeatedly gave "spat" vocalizations. I also heard this vocalization when two males had a dispute over a fallen kily pod and when a female found herself crowded by another to the end of a slender branch. It was usually accompanied by the raised hand. Only once, between the two males, was this actually carried through into a cuff.

The spat call is a series of high squeaks, given in quick succession, usually with a slight facial grin. It is clearly derived with very little change from the infant discomfort call. It is certainly homologous to the spat call of *L.catta,* since it sounds the same, is accompanied by the same facial expression, and is given in the same circumstances, that is, slight aggression. *L.catta's* spat is generally given in defensive threat by the subordinate animal, but I could not tell if this was true for *P.verreauxi.* The spat calls of the two species are easy to distinguish in the wild, for *P.verreauxi's* is so faint that you rarely hear it unless you can see the aggressive encounter, while *L.catta's* shrills through the trees.

TABLE II–10

Propithecus verreauxi; SPATS

Pair	Within Troop	With Single Male
Male-male	3	1
Male-female	–	1
Female-juvenile	1	–
Male-nursing mother.................	3	–
Not identified	2	–
Total spats seen [1]	8	2

[1] Total spats per hour 10/256 = 0.04.

A repertory of aggressive and submissive gestures does appear in territorial behavior, in contact between members of different groups, in threats to other species, and in a mother's warnings to troop members when they approach her infant. In these situations I did not see serious fighting. The aggressive animal may cuff gently with its hand or poke with the end of its closed muzzle. Staring and erect posture, as in all primates, indicate confidence. The submissive gestures are the grin and a hunched posture, expressed fully only by the solitary Sideburn to members of the Red Head troop, but also, at very low intensities, by animals toward mothers with infants.

Propithecus do fight, however. The male's ears are jagged and torn as permanent reminders. Reliable accounts (S. de Guiteaud, personal communication) describe *Propithecus* locked in combat, oblivious to an observer's close approach, with blood streaming over their white fur. Sometimes in play one *Propithecus* grabs another and mouths its ear—a shadow of true fighting.

I do not know whether males within a troop fight during the breeding season for dominance or possession of a (the) female. It is equally likely that at this season territorial disputes become true fights. I saw only one example of contact in a territorial dispute, when the attacking animal made a brief grooming gesture to the animal it caught.

Propithecus remains placid almost all of the year and then erupts into battles which draw blood and may even do serious damage. Andrew (1963*b*) points out that baboons have continuous latent aggression but also have elaborate systems of communication to control such aggression. *Propithecus* is in the opposite situation: it shows no aggression outside the mating season, but at that time it inflicts real damage. In some ways it recalls solitary mammals that ignore, court, or fight each other. But the continuous troop life, the contact and grooming, show that *Propithecus* has simply taken another way of forming societies.

SEXUAL BEHAVIOR

I was not at Berenty for the mating season of *P.verreauxi*. Petter (1962*a*) reports this as the time from January to March for *P.v.coquereli*. It is probably during January and February in Berenty.

It seems very probable that *Propithecus*, like other lemurs, is potentially polyestrus, with a female cycle of 4 to 6 weeks. The *Propithecus* of various troops in a single area breed at the same time. This is deduced from the birth of infants, which was highly uniform among *Propithecus* troops of the region (see p. 65). The mechanism of this synchrony and its social and evolutionary significance, is discussed on pages 155–67.

Petter-Rousseaux (1962) reports gestation time of a captive *P.v.verreauxi* as being from February 20 to August 1, or 130 days. Mating of *P.verreauxi* at Berenty would, from this figure, have occurred about the period of January 25 to February 4.

The one sequence of near sexual behavior that I observed was on March 20, and it culminated only in attempted mounting. The troop's behavior from March 19 to March 21 probably has little relation to behavior in the true breeding season but may serve for comparison when someone describes *Propithecus* mating. This period also gives a sequence of fairly high-intensity social behavior when the females were neither in estrus nor with infants.

On March 19, Rightwhite's vaginal orifice was flushed pink. She was the only female in the Widow Peak troop, which included three adult males: Widowpeak, Widownock, and Tiara, and one juvenile, Faceless. For 10 minutes, from 9:45 to 9:55, Rightwhite and Widowpeak play-wrestle-groomed, facing each other, each attempting to groom the other's genitalia. Rightwhite broke off and jumped to a vertical trunk. Widowpeak leaped on to her back, holding with his feet around her waist in juvenile position (much too high for real mounting) and groomed up the back of her neck, then hopped off. Both joined the rest of the troop, fed, and sunned until their siesta at 12:10. By 3:45 they were feeding again. Rightwhite genital-marked a branch, then leaped beside Widownock. Widownock promptly leaped away from her to smell her mark, with mouth half open and grinning. He smelled again, then urine-marked. The same process happened again 40 minutes later. At 18:15 the troop settled to sleep.

They woke at 5:30 on March 20 and leaped high and freely at 6:05, then settled to sleep again in a locomotive of five animals at 6:20. Tiara turned to groom Faceless, behind him, then hopped to the rear of the line and groomed Widowpeak. The group broke up to different branches at 6:30. Rightwhite sprawled with her feet up, sunning, fed, groomed herself, including her genitalia, and leaped back in front of Widowpeak in a new locomotive.

Widowpeak groomed his own genitalia, with an erection—the only time I saw this in *Propithecus*. He eventually lifted his head, laid it on Rightwhite's shoulder, and subsided. By 7:20 all the animals were asleep. Rightwhite left at 7:30. Widowpeak genital-groomed, then groomed reciprocally with Tiara. The group fed and slept until 12:10, when Widownock again smelled and marked one of Rightwhite's marks.

At 2 P.M. the Dent Ears invaded the Widowpeaks' territory in the first of the series of "battles" described on pages 50 to 52. They were repulsed with much scent-marking, especially on the part of Widowpeak, Widownock, and Rightwhite. Five minutes after the Dent Ears' departure, Rightwhite genital-marked a sloping branch. Widownock leaped to her, smelled, and throat-marked the branch. He followed the smell upward, wiping his throat from side to side on the branch. He ended with his nose at the base of Rightwhite's tail. Rightwhite kept her tail firmly pressed to the branch. Widownock paused for 30 seconds, looking away. Rightwhite leaped off, to Tiara, whom she began to groom. He did not reciprocate; so she left. Widowpeak leaped to Rightwhite and sat facing her, as yesterday, with arms and feet out. He presented his face, put his hand around Rightwhite's nose, and gently groomed her face and head. Rightwhite looked at me, groomed, looked at me, sifakaed once; they both looked at me, and Rightwhite hopped off. The troop began to move and feed.

At 2:50 P.M. Tiara and Rightwhite sat opposite each other in a small bushy tree and began an intense bout of wrestling and mutual genital grooming. Both were constantly active, constantly seeking and changing contacts, but where, in most wrestling, the *Propithecus* simultaneously attempt to groom each other's genitalia, in this, one would lie on its back for several minutes, allowing the other to genital-groom, then the roles were reversed. At 3:25 Rightwhite sprang 2 m away without warning. Tiara followed and clasped her around the waist with his hands, grabbing with his feet for the backs of her legs. The female kicked backwards hard, and both slipped down the tree. As Tiara grabbed for the tree, Rightwhite broke loose. Tiara again half mounted her, although she continued to kick. The female finally left him at 3:45 to feed. The female's attempt to escape, like most *Propithecus* contact, was a slow-motion version of fighting with no chance of danger, unless they fell off the tree. It seemed, though, that the struggle was serious enough to prevent real mating.

On the following day I saw no contact between Rightwhite and any of the males, except 5 seconds grooming with Tiara. By this time Rightwhite's genitalia took a slightly purplish hue—not so pink as before. Through the following months they remained somewhat redder than those of other female *Propithecus,* probably an individual characteristic. Rightwhite did not have an infant that year.

This episode is probably not typical of normal *Propithecus* breeding behavior. It does show that attempted mounting was like that of other primates and that, at least out of season, males may approach the female without apparent rivalry. It also shows that wrestling and grooming, which often occur between males and with little apparent sexual content, also may be part of a sexual relation.

MOTHER-INFANT RELATIONS

In 1963, most infant *Propithecus* in the region were born between July 5 and July 15. This includes five of six born to troops I knew at Berenty, three more in less studied troops, three at Bevala, and very likely one seen at Ifotaka. I did not see the birth of a *Propithecus*, but I saw three infants 1 to 3 days after birth: the two of Door Knob troop and the one of Dent Ear troop. The greatest span between births of which I am certain was between the first Door Knob baby and the Blaze Nose, the first Door Knob was born by July 6, and the Blaze Nose infant not until at least July 13. When Blaze Nose's infant was first seen July 26, it was noticeably smaller than all other known infants, including the three born at Bevala. It seems likely from apparent size that the Red Head and Scrape Tail infants were born between July 5 and July 15, while Blaze Nose's infant was born later. Thus, five of the six infants I could date most accurately were born within 10 days; others in the region were probably born during the same time. (It is perhaps significant that the Blaze Nose juvenile in March, 1963, was fuzzier and more dependent than others; this may mean Blaze Nose habitually gave birth later.)

The newborn infant is perhaps 13 cm long, including tail. Its body reaches only two-thirds across the mother's lower abdomen. Its knees and elbows are bent frogwise, although the arms can extend forward. Its eyes are open, and its head is large as with most newborn infants. Its fur is fairly sparse, so that the black skin shows through, giving the infant a dirty gray appearance. It is easy at this stage to see the direction of hair growth, since the fur still lies limply along the body. The adult's brown cap is only a beige topknot just above the white brow, reminiscent of the upward-growing topknot or crest of *Lemur* species such as *L.macaco rufus* or *L.mongoz coronatus.* Two of the infants at Bevala, born to normal white females, had no topknot at all, but pure white heads like adult *P.v.deckeni,* though the survivor at 9 months had grown a brown cap like any other *P.v.verreauxi.* The infant of a melanotic female was normally patterned white with a beige topknot.

The infant *Propithecus* clings transversely across its mother's lower abdomen in the same position as an infant *L.macaco.* The weight is supported largely by the arms, and fingers and toes are clenched in the mother's fur. The *Propithecus'* tail is slightly prehensile at this stage, although the newborn's tail is only a black, hairless string more suitable for a rat than for a primate. One's first glimpse of a new baby is usually this black string protruding between the mother's thigh and body, wrapped partway around her waist.

In the first few days, the infant moves very little, except to crawl by its arms up to one of the mother's two axillary nipples to suck. If the mother moves, the infant half-scrambles, half-slides down again to her lower abdomen. I have heard of infant *Propithecus* falling and of adults descending to rescue them, but all those I saw maintained a secure hold.

The mother does not help the newborn much with her hands. While resting she takes up the "cuvette" position (Petter 1962*a*). She leans slightly backward, knees up and spread apart, tail curled into a mainspring between her shanks. She rests elbows on knees, or partly blocks the gap between thigh and body with her arms. The result is a living bowl or playpen where the infant can scramble about in safety, protected from falls and practically invisible from the ground.

The mother grooms her infant at frequent intervals, bending down to lick and groom inside the cuvette. The first mother seen, in the Door Knob troop, kept deflecting the gesture to groom her own knees and thighs on either side of the infant. I think this infant was probably only a day old. As the mother grew more self-confident on succeeding days, she usually held the infant while grooming first its head and then its rump, hooked in the large adult hand (see Petter 1962*a*).

The mother makes no concessions to the infant's power of grasp. She feeds upside down and leaps 5–10 m as usual.

By the third or fourth day the infant becomes much more active. It has changed little in appearance beyond having slightly fluffier fur. It will, however, scramble about in the cuvette and often looks over the edge, its black face thrust between the mother's thigh and body. It may even ride with head outside, although it is usually still concealed on the mother's abdomen.

The only vocalization I heard from young infants was a series of soprano click-squeaks like the adult spat but much more faint. They were given as a discomfort call on two occasions when the infant was groomed too hard by a troop member. The call differentiates later into sharp spat-squeaks and into blurred bubbly-squeaks. Juveniles approaching their mother may give blurred bubbly-squeaks as late as February or March, and this call eventually becomes the adult "coo."

Lemur infants in captivity sucked their thumbs. I did not see thumb-sucking in wild *Propithecus*—the infants were usually far too busy holding on (Mason 1965)!

As the infant grows older, it becomes whiter and moves more

adventurously. I first saw the Blaze Nose infant's scrambles take it off its mother onto another troop member when it was about 2 weeks old. The first Door Knob infant, aged 6 weeks, crawled twice to its mother's back and clung there, reaching clumsily for a twig. Eleven days later, the infant left its mother to crawl on a branch but still kept contact with one foot.

On September 4, when the infants were almost 2 months old, I saw one lose complete contact with fur for the first time. An infant at Bevala began to play the game of "leave-mother-and-dash-back," familiar to me in captive *L.macaco*. The infant pulled itself up the small sloping branch where its mother sat and then dropped back onto her. The movements were hilarious: the baby pushed hard with its hind legs on the branch. In an adult this move propels the animal several feet upward, and it catches hold with hands and feet. The infant's uncoordinated legs only bounced its rump an inch or so away from the branch. Then it hauled itself upward hand-over-hand, like climbing on the mother's fur. After two or three bounce-and-hauls, the baby panicked and dropped back, only to start the game again.

At about 3 months, the infant switches to riding on the mother's back. Her belly fur is too sparse to hold its growing weight, and the infant might perhaps impede her leaping. It does not jockey-ride like a young baboon, but remains with ventral surface pressed against the mother, feet grasping her waist on either side and hands on her shoulder fur. The infant, or rather juvenile, rides only intermittently in December and is almost independent by January. The Blaze Nose juvenile did remain close to its mother and often sat near her in the spring of 1962. On March 11, the troop gave a hawk alarm, and the juvenile dropped briefly onto the female's back in the riding position.

TROOP-MOTHER-INFANT RELATIONS

The other members of the troop crowd around a newborn, attempting to groom the infant, or, second best, the mother. The mother remains boxed around the infant, at least at first; so she is actually groomed more frequently. Males and juveniles are as active as females in their efforts to reach the baby. In troops I knew well, every member groomed the infant at some time. Troop members, clustered around the mother, also fell to grooming each other. The only adults which remained aloof were the two mothers in Door Knob troop, who did not show any interest in each other's infants, while the troop divided its time between them. As Table II–11 shows, the relative frequency of grooming is what we would expect if pairs of animals groomed at random, the infant being considered a full member of the troop. Absolute frequency is much higher after the birth of infants.

At first the mother remains somewhat apart with her newborn. Troop members may be discouraged by a lifted paw or even a true cuff. (I saw the submissive grin within a troop only when a male gave it to a mother who had just cuffed him.) Later the mother allows more access to her infant.

TABLE II–11

Propithecus verreauxi; FRIENDLY INTERACTIONS IN TROOPS WITH INFANTS

Pair	Contact	Grooming	Play	Total Observed	Total Expected [2]
Male-male...............	–	5	–	5	6
Male-female.............	2	1	–	3	3
Female-female	–	–	–	0	0
Male-juvenile...........	1	6	2	9	8
Female-juvenile.........	2	–	–	2	2
Male-mother	3	8	1	12	10
Male-infant	1	8	–	9	10
Female-mother..........	1	–	1	1	2
Female-infant...........	–	2	–	2	2
Juvenile-mother.........	1	5	1	7	4
Juvenile-infant..........	–	1	–	1	4
Subtotal	11	36	4	51	51
Mother-infant	–	16	–	16	–
Not identified...........	4	9	4	17	–
Total [1]..............	15	61	8	84	–

[1] Total friendly interactions per hour; 2.0.
[2] Expected on hypothesis that animals in each troop, including infants, came in contact randomly. Calculated for five troops watched.

A typical sequence took place on July 6 in the Door Knob troop, before the second infant was born. The first mother joined a male, Leftline, and slept beside him. She then made a cuvette with her feet up in front of her and her chin on her knees. Three times she groomed the infant. Leftline craned his neck but could not apparently look into the cuvette. The mother groomed Leftline's head, then looked over his shoulder, and he promptly groomed the infant. The mother let him continue for two minutes, then pushed his head away with her nose and closed the cuvette. After a few minutes Leftline pushed his nose toward the mother. She turned sharply, half raised her right hand, and half opened her mouth. Then she shut her mouth and groomed him. Leftline gave two cursory licks to the mother's chest, then groomed the infant. The mother stopped grooming after 20 seconds, butted Leftline with her head, glanced at me, and closed the cuvette.

Another male, Triplepeak, sitting below the mother, groomed the mother's rump above the base of her tail. The mother leaned over, thereby opening the cuvette, while Triplepeak groomed her face. A juvenile, Pyjama, looked into the open cuvette from above while Leftline groomed the infant. The mother kept catching at Triplepeak's hand with her hand; he kept jerking it back and finally ceased to groom her. She straightened up, nudged the other two animals out with her nose, and closed herself around the infant. Both Pyjama and Leftline attempted to shove their noses in to the infant, which gave barely audible soprano squeaks. The mother immediately sprang away from the others. Triplepeak wandered off, eating kily, while Leftline and the juvenile clumped together.

On July 29, Blazenose sat grooming her 2-week-old infant, while Nockear, an old male, sat above eating acacia. Nockear leaped to Blazenose, groomed her first perfunctorily and then hard. At last he groomed the infant, which wriggled and then managed to crawl onto him. Blazenose hopped off to eat acacia. The infant took up a transverse position on Nockear's abdomen, just as it would on its mother. The old male sat for 5 minutes, grooming the infant every few seconds. Then he started to back slowly down the tree. Blazenose leaped over and pawed the infant off onto herself; the infant clutched at both their fur as she scraped it over with her clumsy hand. Blazenose then settled to sun as another (childless) female approached for a chance to groom.

The troop does become rough in its attentions and seems to demand the privilege of grooming; it is not a reciprocal relation.

On August 16, a male grooming a 6-week Door Knob infant groomed the infant's face. The infant reached with its hands and tried to catch the male's face. The male nosed its hands aside. The infant then took the male's hand in both its own hands; the male's wrist was just narrow enough for the infant to hold both sides with arms stretched straight forward. Then, for the first time, the infant itself began to groom with two or three licks or scrapes up the back of the male's hand. The male licked the infant's head and then disengaged his hand to groom harder. The infant scrambled back to its mother's stomach but left one foot outside in the crack of her thigh. The male pulled back on the foot with one hand and licked all the toes. He tugged again hard, until the mother finally nosed him away.

The baby *Propithecus* grows surrounded by doting father, uncles, and siblings. Later, males sit with juveniles less than females do, but the peaceful structure of a *Propithecus* troop starts with this multiple parent relation for the infant. Perhaps this ensures that the infant will grow up as a good family *Propithecus*.

Chapter III

Lemur Catta

GENERAL DESCRIPTION

Lemur catta is the "common" ringtailed lemur (Plate 5). Its fur is a brilliant light gray, its face a black-and-white mask, its tail ringed with about fourteen circles of black and white. Its black skin shows on nose, palms, soles, and genitalia. Your first impression of an *L.catta* troop is a series of tails dangling straight down among the branches like enormous fuzzy striped caterpillars. Later, with difficulty, you put together the patches of light and shade into a set of curved gray backs, of black-and-white spotted faces, of amber eyes. By this time, if the troop does not know you, they are already clicking to each other, and first one and then a chorus begin to mob you with high, outraged barks. The troop is quite willing to click and bark for an hour at a time in the yapping soprano of twenty ill-bred little terriers.

By this time there is ample opportunity to see the details of their builds. *Lemur* walks quadrupedally, its hindquarters standing higher than shoulders or head. It is relatively light, perhaps two-thirds the weight of a similarly sized cat. *Lemur* springs through the branches; so its limbs function as a system of lightweight levers, and it does not need the solid, tearing forequarters of a carnivore. Its head is relatively small in relation to its body; its muzzle is long and pointed. Like *Propithecus*, it carries the muzzle downward, so the *Lemur* can stare directly forward with both amber eyes. It has, however, more the allure of a coati, or even a squirrel, than of a higher primate.

L.catta is fairly distant from the other species of the genus, *L.macaco*, *L.mongoz*, and *L.variegatus*. The first two have brownish fur and similar vocalizations (Andrew 1963*a*), but have a different arrangement of scent glands (Hill 1953) and a number of subspecies. *L.variegatus* is also quite far removed from the others (Petter 1962*a*).

L.catta is peculiar among the *Lemur* because it has no subspecies or polymorphism. It lives in the south and southwest of Madagascar, a fairly large area, and is one of the most familiar of the Lemuroidea. On the west, it overlaps broadly the range of *L.macaco rufus:* both are present in the forest at Lambomakandro. In the east, it meets *L.macaco collaris* in what is now a sharp line, since the western slopes of the mountain range have been entirely burned over, leaving a narrow lifeless division between the "Southern domain" dry forest and the "Eastern domain" mountain forest. The division was always sharp, however, and the transition belt is no more than 15 to 30 km wide, often less. In the north, *L.catta* ranges to the plateau, which is now burned and sterile.

There is no difference in pelage between the sexes. Males have somewhat heavier heads and shoulders than females, but this is difficult to see in the field. The male's scrotum, however, is large, black, and pendulous. If an adult male *L.catta* is walking on a fairly high branch, its genitalia can almost always be seen from ground level. *L.catta* also have a scent gland on the inner surface of the forearm. In males, this is a bare patch of skin perhaps 1.5 cm in width and 2 cm long, surmounted medially by a horny epidermal spine and continued distally in a tract of bare skin to the palm. Montagna (1962) found that this contains an endocrine gland of structure much like testicular Leydig cells. The gland also exudes a secretion onto the arm. It is a very useful characteristic for identifying males in the field.

The adult female genitalia are just below the anus, visible only if the animal is walking directly away with its tail raised. Out of the breeding season, the genitalia are about 1.5 cm long; they are black, and shaped like an elongated pear with the fused vulval lips above and the clitoris below. During the breeding season, the genitalia enlarge to about 3 cm in length and show a brilliant pink vaginal opening (see pages 110–13). Females also have a bare patch of skin on the forearm, about half the size of the males, but there is no horn and only a line in the fur connecting this patch with the palm.

The young are born annually during a brief season; in the Reserve this began during late August in 1963. They are relatively precocious, beinning to reach for branches in the second week. The females nurse as late as March, though the juveniles may have been eating solid food long since. By the next breeding season, in April, the juveniles are about two-thirds the size of the adults. They are still smaller than adults at 1 year, when the following year's young are born.

Subadult animals reach adult size between 1 and 1½ years of

age. Their genitalia, however, are still less than half adult size. Thus in the breeding season, there are 1½-year-old subadult animals which are fully grown but not yet sexually mature. *L.catta* normally breeds at 2½ years of age. In zoos, *Lemur* can live to 20 years of age. There are no data on life span in the wild.

It is relatively difficult to distinguish individual *L.catta*, in spite of their bold face-mask. Males, and occasional females, may have nicked or torn ears. Injuries of the hands, broken fingers or wrists, are also fairly common.

The most striking variation in facial pattern is found in the black eyering. Sometimes it is continuous with the black nose. Some animals can be distinguished by the shape of their white brow. The capline may be concave, straight, convex, or rise to a sharp or blunt point. In a few animals the two sides are asymmetrical. Similarly the line between brow and eyering may be concave, convex, straight, or stepped, and the brow may be wide or narrow (Fig. 1).

I recognized all adult faces only in *L.catta* Troop 1. In this troop three of the five males had visible injuries—two with torn ears and one with a broken finger. Four of the five males had very similar brows: high narrow-peaked capline, white blaze down the center of the brow, and slightly grayer fur near the eyerings. Several males in the adjacent Troop 2 had similar faces. It seems likely that in *L.catta*, as in *P.verreauxi*, animals may show strong family resemblance which makes individual identification even harder.

ECOLOGY

TERRITORY AND HOME RANGE

L.catta, like *L.macaco* (Petter 1962*a*) and *P.verreauxi*, has well-defined territories. The territories and ranges are shown on map VIII. My study area of 10 hectares included the full range of only one troop. This troop in 1963 had twenty animals, and in 1964 had twenty-four. Its total range was 5.7 hectares; the area of large trees in the range was 2.7 hectares. This is an average of 0.3 hectares of range per animal and 0.2 hectares of large trees per animal, which compares with 0.5 and 0.3 per animal for *P.verreauxi* in the same area.

L.catta, like *P.verreauxi*, spends much of its time in large trees and in smaller trees with some clear space above the ground. *L.catta* also rests at midday in liana banks, the animals being just the right size to drape themselves along a knotted liana between 5 and 15 cm in diameter. *L.catta* comes frequently to the ground, parading about in the open spaces under the kilys or feeding on grass and ground plants in the meadows. Like *Propithecus,* the *L.catta* avoid bushes, which have no open stems to climb on or ground space to walk on. Bushes tend to grow on the edge of forest stands where more light penetrates, and there are few tree trunks which offer escape upward. The commonest bush in the Mandrary gallery forest has long, bramble-like stems with recurved thorns. I once saw an *L.catta* catch its tail in these thorns, and although it soon pulled loose, the *Lemur* probably found these plants uncongenial.

A *L.catta* troop, like a *P.verreauxi* troop, follows a daily routine, using one part of the territory for 3 or 4 days and then changing to another part, so that within a week or 10 days, the troop visits all parts of its range. The troop, however, moved relatively farther in each displacement than *P.verreauxi;* so it might cover two-thirds or three-fourths of its range during a day.

BRANCH TYPE

L.catta can climb both large and small, vertical and horizontal, branches. Almost no bark is too smooth to offer handholds; so the lemur can run straight up the trunks of large trees. By preference, though, *L.catta* runs on large sloping or horizontal branches, or climbs saplings small enough for the animal to grasp opposite sides. The animals feed among the smallest twigs, although they do not spread-eagle or hang slothlike beneath the twigs as often as *Propithecus* do. Unlike *Propithecus, Lemur* often rests and sleeps on horizontal branches with no vertical handhold.

Lemur, unlike *Propithecus,* often flings itself onto a bank of foliage rather than aiming for a particular branch. This is common when the troop is moving quickly; one animal after another crashes into leaves and springy twigs, clutches for a handhold, then runs on up a larger branch behind.

L.catta do not apparently choose "safe" branches, but even leap onto dead sticks and loose dead branches lying on live ones. When a branch breaks, the *Lemur* can usually reach or leap to another. I have twice seen animals fall 15 m from a dead branch to the ground. On both occasions the *Lemur* bounced and then ran up the nearest tree without apparent injury. I was told, however, of a mother who fell with her infant; they both were killed (Y. de Guiteaud, personal communication).

BRANCH HEIGHT

L.catta is often called terrestrial (Shaw 1879, Rand 1935, Hill 1953). However, it does not seem to live outside forests, and I cannot picture

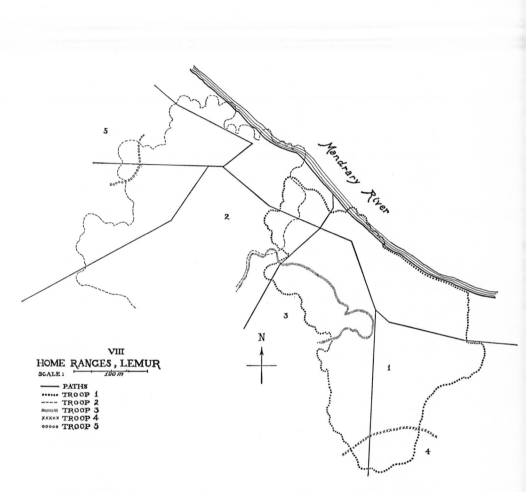

VIII
HOME RANGES, LEMUR
SCALE: |⎯⎯⎯⎯⎯⎯⎯⎯|
              100 m
⎯⎯⎯ PATHS
•••••• TROOP 1
- - - - TROOP 2
⸗⸗⸗⸗⸗ TROOP 3
××××× TROOP 4
ooooo TROOP 5

MAP VIII.—*Lemur* troop home range boundaries. Area the same as
Maps V and VI.

it leaping from rock to rock or foraging through savannah-like open country. It does not even live in the dry, open euphorbia bush, where *P.verreauxi* thrives, although this may be owing to lack of water, not of trees. Hill (1953) says that *L.catta* eats fruits of *Opuntia mona-cantha,* the prickly pear, which was destroyed by the cochineal beetle in the 1920's. It is possible that when the euphorbia forest was full of succulent prickly pears *L.catta* lived there with more nearly terres-trial habits.

Within the gallery forest, *L.catta* spends 15 to 20 per cent of its day on the ground, three or four times as much as *P.verreauxi* (Table III–1) and nearly as much as Uganda forest-living baboons (Rowell, in press). Faint paths through the grass and leaf litter reveal its commonest routes. It is always, however, near trees and ready to spring up the nearest sapling at any scare, even an abrupt movement by a troop member.

TABLE III–1

Lemur catta; HEIGHT ABOVE GROUND

HEIGHT IN METERS	PER CENT OF OBSERVATIONS			
	Mar.–Apr., 1963	June–Sept., 1963	Mar.–Apr., 1964	Avg
0	15	20	15	15
>0–3	10	10	10	10
>3–6	15	10	15	15
>6–12	25	10	25	20
>12–21	30	40	35	35
>21	5	10	0	5

Otherwise it apportions its time between the different heights in about the same way as *P.verreauxi.* Again, this corresponds to the density of vegetable food.

WATER

L.catta lick rain and dew from leaves and in the early morning may do so for 15 or 20 minutes before beginning to feed. They reach into hollows in trees with their hands and lick the drops from their fingers. In the dry season, troops descend to the Mandrary to drink, wherever a half-uprooted tree or straggly bush gives them some cover to reach the water level.

I did not, however, see troops cross strange territory to reach the river. Those troops which did not own any river frontage presumably lived on dew and the moisture in their food.

TABLE III-2

Lemur catta; FOOD PLANTS EATEN

PLANTS	FAMILY	LEAVES, FRUIT, INFLORES-CENCE	PER CENT OBSERVATIONS			
			Mar–Apr., 1963	June–Sept., 1963	Mar–Apr., 1964	Avg
Acacia sp.	Mimosaceae	L	—	1	—	0
Achyranthes aspera	Amaranthaceae	L	4	3	6	4
Albizzia berrieri	Mimosaceae	L	4	—	—	1
Bauhina sp.	Leguminosae	L	—	1	—	0
Canthium sp.	Rubiaceae	F	—	—	3	1
Celtis bifida, C.gomphophylla	Ulmaceae	LF	34	—	24	19
Commicarpus commersonii	Nyctaginaceae	LI	—	18	8	9
Cordia or *Ehretia* sp.	Boraginaceae	L	4	—	1	2
Crataeva greveana	Capparideaceae	L	4	—	—	1
Croton sp.	Euphorbiaceae	L	—	—	1	0
Ficus coculifolia	Moraceae	F	2	—	—	1
Ficus grevei	Moraceae	LF	2	—	1	1
Ficus tiliefolia	Moraceae	LF	2	—	1	1
Mazunta modesta	Apocynaceae	LF	4	3	—	2
Melia azedarach	Meliaceae	L	2	—	—	1
Melothria sp.	Cucurbitaceae	LI	—	1	—	0
Rinorea greveana	Violaceae	LI	2	2	1	2
Solanum sp.	Solanaceae	FI	2	1	—	1
Tamarindus indica	Caesalpiniaceae	LFI	28	68	51	49
Tarenna grevei	Rubiaceae	LF	—	1	—	0
Terminalia mantaly	Combritaceae	L	2	—	—	1
Tricalysia sp.	Rubiaceae	LI	4	—	1	2
?	Graminaceae	L	—	—	2	1
?		L	—	1	—	0
Total			100	100	100	101

FOOD PLANTS

Lemur, like *Propithecus*, does not seem to eat insects or animal food. I once saw an *L.catta* nosing a colony of chrysomelid beetles, but it did not eat any.

I collected twenty-four species of plants eaten by *L.catta* at Berenty, of which only between eleven and fifteen were eaten in any one of the three seasons of observation (Tables III–2, III–3). Of the species eaten in any season, 60 to 80 per cent were also eaten by *P.verreauxi*, and during 75 to 90 per cent of the feeding observations were of these species eaten by both genera.

TABLE III-3

Lemur catta; Food Plants Summary

	Mar.–Apr., 1963	June–Sept., 1963	Mar.–Apr., 1964	Total
(1) Total Species Food Plants	15	11	12	24
(2) Total Observations of Feeding	54	75	166	295
				Avg
(3) Part of Food Plant Eaten (%)[1]				
spp. leaves eaten	85	90	90	90
spp. fruit eaten	45	35	40	40
spp. flowers eaten	25	35	35	30
obs. feeding leaves	25	25	25	25
obs. feeding fruit	70	70	70	70
obs. feeding flowers	5	5	5	5
(4) Plants also Eaten by *P.verreauxi*				
spp. eaten by both *L.catta* and *P.verreauxi* as per cent of (1)	80	65	60	70
obs. *L.catta* eating these spp. as per cent of (2).................	90	75	80	80

[1] Percentages do not add up to 100 because animals often ate more than one part of a species of food plant.

Two ground plants, *Commicarpus commersonii* and *Acyranthes aspera* made up as much as 20 per cent of the *L.catta's* diet during June–September 1963, which accounts for much of the difference from *P.verreauxi*.

L.catta ate the leaves of 90 per cent of its species of food plants in 25 per cent of the observations; in 70 per cent of the observations it ate the fruit of 40 per cent of the species. As with *P.verreauxi*, kily pods were the year-round staple. In 1964, when there had been relatively less rain during the wet season than in 1963 and two cyclones had shaken down the fruit, *L.catta* fed largely on old kily pods from the ground under the trees.

There is no comparable data from other forests. However, *L.catta* also lives in the forest of Lambomakandro, where the available food presumably differs from that at Berenty. In captivity, *L.catta* accepts a wide range of natural and artificial food, from parsnips to chocolate ice cream.

L.macaco at Ankarafantsika, in northwestern Madagascar, also feeds on kily pods, and in Nosy-bé, a northern island, also eats species of *Ficus, Terminalia,* and *Albizzia* (Petter 1962*a*). A single *L.macaco collaris* from the mountain rain forest seemed to thrive on the diet of *L.catta*. It seems unlikely that *Lemur* species are separated by any particular food requirements but can live on the plants of almost any Malagasy forest.

INDIVIDUAL BEHAVIOR

ACTIVITY

L.catta is far more active than *P.verreauxi*. The troop may move 200 m in a single displacement and would ordinarily travel a route of 500 to 600 m in a day. Individual animals moved much farther, leaping back and forth in the branches or wandering in a zigzag course on the ground. However, they also rest for long periods during the day and are not active by comparison with the livelier Old or New World monkeys.

In general, a troop of *L.catta* wakes before dawn and moves about in the branches. It starts to move, sun, and feed at any time between 5:30 and 8:30 A.M., depending on temperature and weather conditions. There may be two long progressions in the early morning, one just after the troop wakes and one between 8 and 10 A.M., when the troop moves from the high trees to low trees or to the ground to feed. Another progression about noon takes them to the site of their siesta. The troop may change position again in early afternoon and again settle or may remain in one place until the late afternoon feeding period. A long progression just after sunset brings them to the sleeping trees. The troop as a whole does not change trees during the night, but animals leap, feed, groom, and spat during the hours of darkness.

A troop may follow about the same course during 3 or 4 successive days; it then usually shifts to another part of its range.

MANIPULATION OF OBJECTS

Lemur, including *L.catta*, does poorly on primate "intelligence" tests (Klüver 1933, Bierens de Haan 1930, Jolly, 1964 *a, b*). It has little understanding of the relations of strange objects and learns with difficulty to look in containers for hidden food, to pull in food on strings, or even to reach with its hand into small bottles to pull out a reward. *Lemur* ordinarily attacks such a problem with nose and mouth while hanging onto the object with both hands.

In the wild, *Lemur* draws in a branch with its hand and seizes the fruit or leaf with its mouth. It does fine manipulation of objects with the mouth, prying seeds loose with the tooth-scraper or cracking pods with the molars. Thus, *Lemur* treats the objects in psychological tests just as it would a piece of food.

In one test (Jolly 1964*a*) jingling hardware puzzles were put in six cages of *Lemur*. In each of four cages, a pair of *Lemur* played with the puzzle for 10 days.

One adult male *L.catta,* caged alone, ignored the puzzle and other proffered toys completely. This was a rather fat and lethargic animal, surprisingly affectionate toward human beings for a male *L.catta*. He had been born and raised in captivity and was caged for almost 6 months with another male; he then had 6-months solitary confinement before his psychological testing. His resulting character resembled that of subordinate males in normal troops which (out of breeding season) trailed behind the troops, groomed each other, and showed practically no aggression or other social initiative toward troop members.

The sixth cage contained two male and two female *L.macaco* in a stable social pattern. They were constantly interacting with each other and inventing locomotor play in their large enclosure, but they ignored the puzzle. One of the females, who was tame to me, would, however, play with any object I offered her personally; she would pull, kick, lick, and chew it.

In the wild I did not see *L.catta* play with or investigate any inanimate object. They only picked up or handled pieces of food. Petter (1962*a*) reports the same for *L.macaco*. On the other hand, they had a wide repertory of locomotor and social play. On three occasions *L.catta* approached and investigated strange species: once a beetle, once a *Lepilemur*, once a human being. Teasing *P.verreauxi* was standard sport. Therefore, both in cages and in the wild, where there is opportunity for social and locomotor play, *Lemur* seems to ignore inanimate objects.

I have suggested (Bishop 1962, Jolly, 1964*a*) that a tendency to play with objects might have been a pre-adaptation to greater intelligence and understanding of objects. The tendency is certainly present in *Lemur* but depends for expression on the social and environmental conditions. *Lemur* in partly adequate groups in barren cages may turn their attention to objects, but in wild groups, their attention is directed elsewhere. The same is true of captive baboons when compared to wild ones (T. Rowell, personal communication).

SENSES

The eyesight of *L.catta* is not so acute as that of *Propithecus*, or else the *L.catta* pays less attention to visual detail. I have on several occasions escaped notice while sitting motionless in full view, until the *Lemur* were between 3 and 4 m from me. Then one would start and stare, alerting the others. They never failed to notice me if I moved. Like *Propithecus*, *Lemur* hold their muzzles downward and are themselves brightly colored; therefore they probably have stereoscopic color vision.

L.catta probably see well at night. They leap among the branches even on moonless nights. They often did not reach the sleeping trees until an hour after sunset and moved restlessly after dark and before dawn. On nights of partial moonlight, when the path was dappled with black and silver, I could move only one step every 15 seconds; otherwise one troop member would see me and begin to click or bark. A cat that glided across the trail in flecked moonlight drew the full barking chorus of the *L.catta*. *L.catta* itself is clearly visible at night. Striped tail-rings are one of the most unequivocal markings in dim light, since they can be recognized whatever the animals' orientation. Both starlight and moonlight are bright in Madagascar and the forest is quite open; so the *L.catta* may need only slightly better night vision than the human dark-adapted eye.

L.catta has many vocalizations, including a few which carry (to the human ear) for 750 to 1000 m (the bark, howl, and scream) and one (the purr) which carries about 1 m. They did not seem to hear the approach of other animals either more or less quickly than I.

Scent is obviously of great importance to *L.catta*. Food such as kily pods are brought first to the nose and then, if not rotten, eaten. An animal may also nose over the ground in search of fallen pods. Scent displays, such as tail waving, force an individual animal's odor to another's attention, and marking the branches leaves the odor in the territory, where other *Lemurs* will encounter it. I could not, however, smell these individual scents, even at a meter's distance (in contrast to that of *L.macaco*).

SLEEP AND NIGHT ACTIVITY

A troop of *L.catta* sleeps in two or three large adjacent trees, usually kilys. The troop rarely slept in the same trees on successive nights, although there were a few usual sleeping places. The long troop progression just after sunset could bring it from its feeding area to almost any group of sleeping trees.

The troop could be moderately active or quiet at almost any hour of the night, whether or not there was a moon. I made too few night observations to tell if there was a usual rhythm of activity with, perhaps, more movement before midnight and a quiet period afterward corresponding to the daytime siesta. Neither could I be sure that there was more activity on moonlit nights, although the moonlight made it possible to see the black-and-white tail-rings even at half-moon.

During the night, the animals walked along branches, occasionally leaped between them with a crash, and cracked and dropped kily pods. They also had spats with each other and during the breeding season would howl every hour or two throughout the night.

A *L.catta* might sleep either curled up in a crotch; sitting holding a vertical branch in front, with its feet up and chin on its knees; or stretched out along a branch. There are any number of possible positions with spine multiply curved and feet and hands splayed in unlikely directions. Only once, during a siesta, did I see an animal lie on its side, curled in a circle like a cat, on a very broad branch. In general the weight rests either on haunches or stomach. During siestas, and very likely at night, animals often sleep in contact.

The troop becomes more active, with more movement and spats, in the hour or half-hour before dawn, but it may quiet down as the first light appears. The troop may leave the sleeping trees as early as 5:30 A.M., or in colder weather or in mist or rain, as late as 8:30 or 9:00 A.M. The following displacement may be 10 m to the next group of trees or 300 m to the other side of the territory. Occasionally the troop suns in the same trees where it slept.

LOCOMOTION

Lemur usually runs quadrupedally on the tops of branches or leaps between them. In the leap *Lemur* is propelled by the legs; in midair it has the body tilted forward and arms up like a jumping man; it lands with hind legs brought forward to touch first, like a *Propithecus.* *Lemur* can also land with all four hands and feet clutching a foliage bank.

On branches, *L.catta* very rarely hangs underneath by feet or hands or progresses hanging from all fours. I saw an *L.catta* brachiate only once, in play.

L.catta walks on the ground but may speed up to a gallop, springing with the hind legs and landing on first one, then the other hand. *L.catta* who are disturbed on the ground rise on their haunches to stare about and may hold this pose for several seconds. Then they hop off on their feet, body upright like *Propithecus,* and hop up the nearest tree. *L.catta*'s fastest ground locomotion is a straight galloping dash, when the animal has 5 or 7 m in which to gather momentum. If there are only 3 m to cover, though, the animal reaches a tree as quickly by hopping.

L.catta's tail balances its leaps in the trees. In ordinary arboreal climbing or sitting, the tail dangles straight down. If the animal stands perpendicularly across a small horizontal branch, most of the weight centers on the hind feet, and the head and shoulders are somewhat balanced by the tail. On the ground, the tail rises upward in the shape of a jaunty question mark.

SUNNING

L.catta, like *P.verreauxi* and all other diurnal lemurs, spread themselves in the early sun (Plate 6). The *Lemur* climb to the upper, outer branches of a tree which faces east. Typically they sit on their haunches, arms outstretched to either side, hands dangling limply from the wrist. Their heads also loll to one side, eyes half-closed. Their white belly fur gleams, and the tree is dotted with the repeated white "praying" shape. Occasionally an animal turns around to warm its back in the same position. *L.catta* groom and also feed during the sunning period. They are perhaps more likely to feed than *P.verreauxi,* because *P.verreauxi* often suns on bare branches, while *L.catta* tends to sun on the outer twigs of kilys.

SELF-GROOMING

At intervals, in the sunning trees, animals complete their morning toilet. They lick and tooth-scrape their back fur with the repeated, unmistakable forward movement of head and muzzle. They bend down and scrape their belly fur and lick their palms and their thighs. They groom their genitalia and finish with a scratch with the specialized toilet claw on the second toe or use the toe to clean their ears (Plate 7). The grooming gestures are relatively standard among lemurs, and those of *L.catta* resemble those of *P.verreauxi.*

URINATION AND DEFECATION

L.catta does not scent-mark with urine or feces. An animal simply stands on a horizontal branch, lifts the base of its tail, and lets the urine or feces fall to the ground.

When the troop moves after resting, as after sunning or the siesta, many of the animals urinate and defecate. It is common for one animal to defecate from a branch and for others to follow suit as they reach that branch. If the troop is slightly suspicious of an observer, they may choose to cross directly above the observer and defecate either near or on him. The observer then tends to change position rapidly, but the *L.catta* continue to defecate from the same branch.

The feces vary in nature with the season, being either green or brown, hard or soft. They usually contain quantities of undigested kily seeds. It is likely that *L.catta*, like *L.macaco* (Petter 1962a) and *P.verreauxi* digests the thin seed covering while passing the seed itself.

FEEDING

After the progression *Lemur* spend the second part of the morning in feeding. They may feed in large or small trees or on the ground. Perhaps the most typical pattern is to feed for an hour in small trees, then descend to earth and browse until noon, but there are many variations.

In the large kily trees, *L.catta*, like *Propithecus*, reaches out for a pod, or the pod-bearing twig, and draws it in to its nose without any attempt to break off the pod. It runs its nose up and down the pod, and often then lets the twig go. The kily seeds were edible to lemurs in June, although they did not ripen until August. They continued to be a staple food until the following April, when most had been bored out by insects. The *Lemur* might reject five or six before biting one. Juveniles sometimes bit a dry, eaten-out pod and dropped it, mouthing the mouthful with exaggerated chewing gestures (see pp. 101–4). An accepted pod is held in the hand, one end crunched between the molars, and a seed swallowed. The lemur eats two to four of the seeds and almost always drops the rest of the pod.

During March and April of 1964, the *L.catta* ate many fallen kily pods from the ground, probably shaken down by the cyclones of late February. They found these by sniffing over the litter under a kily tree. They often sat on their haunches to eat, like a squirrel, leaving both hands free to hold the pod. The pulp of these old kily pods had fermented slightly, tingling on the (human) tongue like just-turned cider. The *Lemur* sat licking this pulp rather than just cracking and swallowing the seeds. Most of the *Lemur* spats I saw over food (though there were many in other contexts) were when an animal took one of these old kilys away from another. *L.catta* ate kily flowers, which may grow right out of the bark and may be snapped up in passing.

In large and small trees *L.catta* also ate leaves and small berries by drawing in a twig with one hand or by working its way out to the

PLATE 1.—Aerial view of the reserve.
(*Courtesy of the Service Geographique de Madagascar.*)

PLATE 2.—The Mandrary River from the reserve.

(*Photo C. H. Fraser Rowell.*)

PLATE 3.—*Propithecus verreauxi verreauxi.*

(*Photo C. H. Fraser Rowell.*)

PLATE 4.—An adult female mutual grooms a bald-thighed male in the reserve.

PLATE 5.—*Lemur catta* eating tamarind pod.

(*Photo C. H. Fraser Rowell.*)

PLATE 6.—Sunning.

(Photo C. H. Fraser Rowell.)

PLATE 7.—Scratching with the toilet claw.

(Photo C. H. Fraser Rowell.)

PLATE 8.—Two females and a juvenile on a log.

(*Photo C. H. Fraser Rowell.*)

PLATE 9.—Contact during siesta.

PLATE 10.—A spat in which the inferior male did not notice the superior's approach, then leaped hastily away.

(*Photo C. H. Fraser Rowell.*)

PLATE 11.—Troop progression. Adult females at extreme left and right with light-centered genitalia. Subadult female (1½ years) second from left, juvenile (7 month) third from left.

(*Photo C. H. Fraser Rowell.*)

PLATE 12.—*Lemur macaco collaris* male.

(*Photo C. H. Fraser Rowell.*)

PLATE 13.—*Lemur macaco* with Troop I *Lemur catta.*

(*Photo C. H. Fraser Rowell.*)

PLATE 14.—*Lemur catta* female cuffs the *Lemur macaco*.
(*Photo C. H. Fraser Rowell.*)

PLATE 15.—Male *Lemur catta* wrist-marks tail with flared lip and ears back.

PLATE 16.—Female *Lemur catta* genital marks twig.
(*Photo C. H. Fraser Rowell.*)

PLATE 17.—Male *Lemur catta* in open-mouthed threat with stare.
(*Photo C. H. Fraser Rowell.*)

PLATE 18.—Male *Lemur catta* howling with pursed mouth.
(*Photo C. H. Fraser Rowell.*)

PLATE 19.—Melanotic *Propithecus verreauxi verreauxi*.

PLATE 20.—*Lemur catta* juvenile grooms its mother.

finest twigs. Leaves and berries were nibbled directly from the branch. On the ground *L.catta* followed the same system as in the trees, either browsing directly or pulling the stem of a plant with its hands to bring the tenderer leaves within reach.

SIESTA

L.catta sleeps in the hot season from about noon to 4 P.M. and for a shorter time in the cold season. Unlike *P.verreauxi*, a troop is quite likely to move during the siesta, either in a long progression when the entire troop displaces itself, or in a series of short driftings.

 L.catta sleeps, not in the bushy foliage of a small tree, but on the branches of a large tree or in a tangled bank of lianas. Each animal chooses a section of bare stem, horizontal or sloping, and sits or stretches out on it, hands and feet holding other branches at unlikely angles. An animal may sleep for an hour without moving; it then rearranges itself or changes branch and sleeps again.

 During a siesta animals may sleep in contact, groom, or self-groom. There are few spats (Plates 8 and 9).

EVENING FEEDING AND SETTLING

Between 3 and 5 P.M., the troop wakes and progresses from the siesta site to the evening feeding site. At evening, the troop almost always feeds both in trees and on the ground, moving from one to the other. In the dry season, it may descend to the river to drink, well under cover of bushes, reeds, or a half-fallen tree. There are many social interactions, both grooming and spats. Animals may play, either alone or with each other, and leap about from tree to tree. There is a second long progression, often with a parade up a path, when the troop walks as a group, tails in the air; it does not stop to feed during this movement. The progression from feeding spot to sleeping trees is usually the longest of the day. The troop does not reach its sleeping trees until the last light is fading, ½ to 1 hour after sunset, between 6:30 and 7:30 P.M. I could not be sure when or how the troop finally sleeps, for a few leaps and crashes continued long after it was too dark for me to see.

RELATIONS WITH OTHER SPECIES

HAWKS

L.catta, like *P.verreauxi*, has aerial predators, the hawks, ground annoyances, dogs and human beings, and arboreal companions, the many species which share the trees.

A *L.catta* troop shrieks at the silhouette of a low-flying hawk. The shriek is a high pitched, loud call, resembling a human scream to the ear and given with the corners of the mouth drawn fully back. A terrified *L.catta* that is being seized by a human captor will also shriek (Andrew 1963*a*). As in other *L.catta* calls, animals of a troop shriek simultaneously. In general *L.catta* troops do not shriek at high-flying hawks and rarely at airplanes, although they look up nervously, especially when nearby *P.verreauxi* roar. They may also stream lower in the trees, well under cover, at the sound of a roar.

I once saw a hunting *Gymnogenys,* the great gray harrier hawk, make six passes over an *L.catta* troop. When the hawk first flew to a dead limb above the troop, the troop meowed (contact call) and moved downward into thick bushes and small trees. On the first pass, not very low, over the troop, the troop again meowed. On the next five, the troop shrieked each time and adjacent *P.verreauxi* roared. The hawk did not descend, but simply skimmed the bushes containing the *L.catta.* The *Lemur* remained hidden for some minutes after the hawk's departure. Boggess and Smith (personal communication) saw a brown hawk of another kind make passes over adjacent troops of *L.catta* and *L.m.rufus* at Lambomakandro. This species of hawk later flushed and killed a guinea hen.

An adult *L.catta* could certainly offer some resistance to *Gymnogenys,* but the hawk would be able to lift and kill it. Both the *Gymnogenys* and the brown hawk were making passes over troops of adults and half to three-quarter grown juveniles. An infant lemur would easily be taken.

HUMAN BEINGS AND GROUND PREDATORS

I saw no native carnivores in the Reserve, but *L.catta* has a well-developed series of warning calls which it gives for ground predators. They were directed at human beings, dogs, and cats. When one *L.catta* began a warning, all others left the ground or climbed higher in bushes or low trees.

An *L.catta* which first sights a potential predator stares at the intruder with head usually held lower than the shoulders. The *L.catta* begins a series of single clicks, about two per second. Often it swings its dangling tail from side to side in the same rhythm, like a pendulum (this gesture is even more common in *L.macaco* and in *Hapalemur*). As this clicking or ticking rises in intensity, it becomes a series of click-grunts—short bursts of three or four blurred clicks, still given about two per second (Andrew 1963*a*). The click-grunts grow louder and more frequent. By this time other animals may also be clicking or click-grunting. It often happens, however, that one or two suspicious animals stare and click for 10 or 15 minutes while the rest of the troop tries to sleep. Occasionally the clicker subsides, but even after 10 minutes it may rouse the rest of the troop to mobbing.

Then one animal begins to bark. This is a high-pitched call, given in short bursts of three to eight individual sounds with about five bursts following each other in quick succession (spectrograph in Andrew 1963*a*). The corners of the mouth are drawn back. The animals bark in perfect synchrony; so as soon as one animal gives the first bark in a burst, the rest of the troop joins in the burst. The troop will look at a clicking or grunting animal and stare where it is staring, toward the intruder; therefore the troop's attention is directed toward any predator. However, barking involves the others, whether or not they have seen the source of disturbance. Those which see the predator bark directly toward it; the others yap anyhow. On one occasion an animal clicked, grunted, and finally barked at the sound of my footsteps. I left, sure that at most one and probably none of the troop saw me. The troop continued to bark for 12 minutes after my departure.

Andrew (1964*a*) has commented on the convergence of vocalizations in cats, dogs, and *L.catta*. They purr, growl, bark, wail, and howl, and do so in roughly similar situations. The convergence is startling when several dogs are mobbed by a troop of *L.catta*. The two species approach with click-grunts and growls; the dogs put their forepaws on the trees and jump; the *L.catta* descend to branches just out of reach; and both sides bark themselves hoarse.

SNAKES

There are no true poisonous snakes in Madagascar. The Madagascar boa constrictors eat small mammals, but they are mainly ground-living and probably pose little threat to lemuroids. The one confrontation that I saw between lemuroids and a snake at least shows that *L.catta* are not instinctively afraid of serpents.

A *L.catta* troop had settled for a siesta at 10:40 A.M. in the low open branches of a kily and its dependent lianas. About 10:50 a restless juvenile began to click and stare toward the leaf litter at the root of some lianas. Two other *Lemur* were sleeping in these lianas about 15 feet up. The juvenile approached and descended the lianas to 10 feet, still clicking. After 3 minutes it persuaded two females to follow it. The females clicked a little; the juvenile stopped clicking and stared. At 11:00 a common brickwork snake appeared, about 4 feet long, crawling slowly through the leaves. I could see fourteen *Lemur*, of which three watched the snake, two groomed themselves, and the rest dozed. One female descended to 4 feet above the snake, weaving her head back and forth as she watched it but no longer clicking. The snake crossed the entire clearing under the *Lemur*'s tree, reaching the other side at 11:25—but after 11:05 all the *L.catta* were asleep.

BIRDS AND BATS

L.catta, like *P.verreauxi*, avoided the roosting colony of the flying fox, *Pteropus rufus*. The bats again ate the same food as the lemurs. *L.catta* is still active and feeding at dusk, when the first bats fly; and occasionally a bat landed in the upper branches of a tree full of feeding *Lemur*. On one very windy day the bats were shaken out of their trees and wheeling over the forest in daylight. Many landed in a liana bank that had a *L.catta* troop at the foot. The *Lemur* continually glanced up but did not take cover or move away.

The gray parrot, *Coracopsis vasa*, fed and perched in the same trees with *L.catta*. In general the parrot avoided close proximity with the lemurs, and the lemurs ignored the parrot. I saw no interaction of *L.catta* with other birds.

BEETLES

I once saw *L.catta* investigate colonies of a Chrysomelid beetle, probably *Protogaleruca constulata*.

The chrysomelid may have had an unpleasant scent (Andrew, personal communication).

The colonies were on the bark of a *Rinorea greveana* in mats about 30 cm across. The lower layer of the mat was composed of black pupae, while newly emerged, soft yellowish adults clung to the surface of the mat, along with green harder adults. An adult *L.catta* clung to the side of the tree, nosing the adult beetles without disturbing the pupae. The *Lemur* continually licked its nose and made exaggerated chewing motions, a common displacement movement particularly during aggressive scent-marking. It did not take any beetles into its mouth. Three times the *Lemur* reached with its hand toward the end of the cluster, hesitated, and then held its hand stiffly in midair while reaching further with its nose. At last most of the adults were nosed off and had fallen to the ground. Another *Lemur* sat by another colony and sniffed and nearly touched the colony with its muzzle, but this colony was not disturbed. Both these lemurs and a third one descended to the ground and sniffed at the fallen beetles.

Propithecus verreauxi

Interactions between the two genera are described in detail on pages 43 to 45. In brief, *L.catta* and *P.verreauxi* were both in my sight and in sight of each other for 80 of the 400 hours I watched *L.catta*. They almost always ignored each other, even at a distance of 2 or 3 m. Occasionally *L.catta* actively teased the *P.verreauxi*, and very occasionally *P.verreauxi* supplanted or teased the *L.catta*.

Lepilemur mustelinus

Lepilemur mustelinus leucopus is a small nocturnal lemuroid about half the size of a *Lemur*. It lives in both euphorbia and gallery forest. It ordinarily sleeps in hollows in trees (Petter 1962*a*, H. de Heaulme, personal communication), but I have seen it in both forests sleeping in an open crotch. One *Lepilemur* slept quadrupedally on an open branch of the gallery forest. It did not wake or move as spots of sun passed over it. It woke, but did not move, when eight successive *L.catta* crashed onto and off the next branch.

One *L.catta* then bounced to the *Lepilemur's* branch and walked to within 0.5 m. of it, nose forward. The *Lepilemur* simply swiveled its head a slow 180° and stared down its back at the approaching *L.catta* with round, orange eyes. The *L.catta* checked, hesitated, walked backward hastily for three or four steps, and leaped off after its troop.

TROOP STRUCTURE AND INTERTROOP BEHAVIOR

COMPOSITION OF TROOPS

The composition of three troops is shown in Table III–4. The troops range from twelve to twenty-four adults and juveniles. *L.catta* are fairly difficult to count, since they rarely move in single file; therefore data is most accurate for Troop 1, where I knew individuals.

TABLE III–4

Lemur catta; COMPOSITION OF TROOPS

Troop	Males	Females	Subadult Males	Subadult Females	Juveniles	Infants	Total
September, 1963							
1 [1]	6	9	1	–	3	(4)	20
2	8	3	1	1	3	(2)	16
3	4	7	–	–	1	?	12
April, 1964							
1 [1]	5	9	1	1	7	–	24
2	8	5	2	1	5	–	21

[1] Troop 1 also had one adult male *L.macaco collaris.*

In 1963 the adult male–female ratios in troops 1, 2, and 3 were 0.7, 2.7, and 0.6. The ratios of troops 1 and 2 in 1964 were 0.6 and 1.6. The sex ratio clearly varies among neighboring *L.catta* troops. They do not apparently share the strong tendency of *P.verreauxi* and *L.macaco* to have a male–female ratio greater than one.

There were six juveniles in troops 1 and 2 in 1963, and five subadults in 1964. I could not count all infants born in 1963. In

1964, there were fourteen adult females in troops 1 and 2 and twelve juveniles. Presumably each female has an infant in most years. Twins occur, but rarely, in *L.catta* and other *Lemur* (Hill 1953).

An adult male *L.m.collaris* was a permanent member of troop 1. His behavior, and that of the troop toward him, is discussed in chapter IV.

INTERTROOP BEHAVIOR

I saw only four encounters between troops 1 and 2 and one between troops 1 and 4. I may have seen other meetings of troops, but *L.catta* does not have simple identifiable behavior of territorial defense. If I saw a *Propithecus* troop bunched together, leaping forward rapidly, not in single file but all together, I was fairly certain it was a territorial battle. The only way to be sure two *L.catta* troops are meeting is to see them approach from different directions and to know the individuals.

On March 28, 1964, troop 2 members saw troop 1 ahead in a liana bank and clearing they both used. Troop 2 changed course and did not mix with troop 1.

On April 8, troop 1 was in the same clearing when troop 2 approached. The males of troop 1 began a series of palmar markings, spats, and chases *within* the troop, as though either redirecting or "showing off" to troop 2. The males then retreated to the east side of the clearing without directly confronting members of troop 2. Troop 2 occupied the west side. One female of troop 2, who had a bitten ear (unusual in females; so she was named Kate the Shrew), leaped across to a sapling in the clearing with three of troop 1, who scrambled out to the east. Both troops then retreated slightly, troop 2 leaving after 15 minutes, troop 1 after 45 minutes. On April 11, the troops again met on opposite sides of this clearing, again with intratroop spatting but involving females this time. Both troops retreated.

By the next week, both troops were breeding. On the morning of April 23, Cataract, a female of troop 1, was mating with Jagear, a male of troop 1, who also was chasing off Bunthorne, a persistent rival male of the same troop. Males of troop 2 at the same time attempted to mount and mate with females of troop 2, although successful mating was not seen until that afternoon (Observations by T. E. and C. H. F. Rowell). Cataract and Jagear mated successfully slightly east of the main troop, between troops 1 and 2. Between mountings, Jagear chased Bunthorne, the rival. Then two males of troop 2 joined the chase, which doubled back toward troop 1. One troop 2 male tail waved to Bunthorne, a usual challenge between males of the same troop. Bunthorne faced him, and both animals chewed and yawned 2 m apart. All four males began to howl. I do not know which howled first. One troop 2 male smelled and palmar-marked at length the portion of liana where Jagear and Cataract had mated. The four males howled again, and males within troop 2 howled at a distance. The two troop 2 males started toward their own troop after a brief spat between themselves. Bunthorne, Jagear, and one troop 2 male settled into postures of sleep after howling. Cataract, the mating female, moved down a branch toward Bunthorne. Jagear palmar-marked three times and marked his tail, staring toward Bunthorne and Cataract. The remaining troop 2 male approached, palmar-marked, and marked his tail. Jagear squealed toward this male, who marked again; whereupon Jagear gave the defensive spat vocalization. The troop 2 male approached Cataract, but Cataract chased him. Jagear then pursued the male, who gave the defensive spat call but stopped to tail wave. Jagear jumped on the male, and both fell together. The male ran off a little, marked, and waved his tail. A subadult male, a female, and a juvenile all chased the strange male, who finally moved off to his own troop.

Thus, it may well be possible for females to mate with males of adjoining troops. In general, though, the troops do not mix. When troops meet, individuals approach each other with ordinary intra-troop aggressive gestures, not specially ritualized forms of hostility.

As with *P.verreauxi*, there are likely means for gene exchange between neighboring troops, but it is unlikely that animals move as much as two troops away.

SOCIAL BEHAVIOR

TROOP SPACING AND CONTACT CALLS

When a troop of *L.catta* progresses from one place to another, the females, juveniles, and dominant males move first. Subordinate adult males lag behind (see also Bolwig 1960). The troop spreads over 20 to 30 m from beginning to end. When the troop stops, the subordi-nate males, the "Drones' Club," are thus at one end; they also tend to feed and take siesta near each other.

The troop is generally spread over several trees at several heights, although most of the troop is usually together at one level of foliage. They sleep in three or four adjacent large trees.

During progressions, the troop very rarely doubles back on its tracks so that the head crosses through the tail. It moves instead in wide circles through the range.

No one animal consistently leads the others or initiates progres-sions. At any time the leader is likely to be a female or dominant male, since they are usually in the front of the group. Occasionally the group splits into two factions which start in divergent directions. The factions then mew loudly to each other until one changes direc-tion and joins the other.

Individuals do not keep any standard distance from each other, or rather, the distance varies with any pair of animals in any situa-tion. In general the distances are greater between dominant males than between females. Animals may feed a meter or less from each other.

Before a troop starts to move, the animals give single clicks and

click series. Often there is a bout of moans and wails before movement starts, it begins with gentle moans and are answered by high-
intensity wails. The moan is simply one syllable, given with corners
of the mouth partially drawn back and mouth open: "mew" (Tables
III–5 and III–6). In the wail, a second syllable is added with lips
pursed forward and pitch usually dropped: "me-ow." At very high
intensities, as for instance, when given by an animal out of sight of
the troop, pitch may rise on the second syllable: "me-ow!" Occasionally this second syllable becomes so explosive as to merge with a
bark. All the meow group of calls are answered in kind; in general,
the louder and more intense the call, the more animals answer the
more loudly. The meows certainly function as contact calls within
the group.

Occasionally when I left the troop after prolonged observation,
the animals gave a soft meow chorus. Captive rhesus make a similar
"group contact" call on the departure of a familiar observer (T.
Rowell, personal communication).

One call which I find difficult to analyze is the "howl" or "song."
It is included here because it does in part function as a long-distance
contact call, although this is far from its full explanation. The howl
is similar to the meow calls—a loud clear call in two syllables with a
yodeler's break (up or down in pitch) between. On the second syllable, the head is thrown back nearly vertically with pursed mouth
pointing upward so that mouth and throat increase the resonance. To
the ear, *L.catta*'s howl resembles the song of *Indri*.

A howl can be heard 1 km, which means in the Reserve that an
animal howling in a centrally located troop could be heard by at least
eight or ten other troops—the four immediately surrounding the
howler and the next ring of more distant troops. Howls are almost
always given by males. I did see a howling female once, but I now
almost doubt the identification, since I could never repeat this observation.

Usually at dusk, when the troops settle, one animal of each
troop howls. Sometimes also there is howling at noon, before the
siesta. Individual males are sometimes seen to howl, then drop immediately to sleep.

During the breeding season, males howl far more frequently
throughout the day and at perhaps hourly intervals from dusk to
midnight. Not only will one male howl, but after one starts, others
may also howl, with an effect very like *Indri*, where first one, then
another, then another of the troop give song. During the breeding
season, howls in one *L.catta* troop may be immediately followed by
howls in another. If more than one male howls at once in a troop, the

TABLE III-5

Lemur catta; VISUAL AND OLFACTORY DISPLAYS

Sign	Accompanying Vocalization	Spp. with Homologous Signs	Occurrence
Stare	—	all lemuroidea	mild threat, to unusual objects, to predators
Eyelids lowered	—	—	when stared at, when sleepy
Lips pursed forward	—, meow, howl	I.indri, P.verreauxi	with bouts of chewing and yawning, always with meow and howl
Open mouth, teeth covered	bark	—	mobbing ground predator
Grin	—, spat, meow	P.verreauxi, L.macaco	with play-bite, with all spat calls, in first syllable of meow
Flared lip	—, purr, squeal	—	while marking tail, tail-waving
Chewing	—	—	while scent-marking, in stink fight, rarely to observer or strange object, juv. biting old kily pod during stink-fight
Yawning	—	—	with stare, in all aggressive encounters, while tail-marking, while tail-waving
Ears flattened	—, spat	L.macaco	
Stand on hindlegs	—	—	investigating on ground, mild alarm on ground, in jump-fights
Swagger	—	L.macaco	superior male toward inferior
Stiff-legged leaps	—	—	juveniles in play
Tail pendulum swing	—, clicks	L.macaco, Hapalemur	with stare at ground predator
Touch noses	—	P.verreauxi, L.macaco	greeting
Groom	—	all lemuroidea?	between any two animals
Nose-poke	—	P.verreauxi, L.macaco	female dislodging juvenile, preventing animal from grooming infant
Cuff	—, spat	P.verreauxi, L.macaco	in threat, spats, jump-fights against P.verreauxi
Feint with shoulders	—	—	against P.verreauxi
Bite	—	all lemuroidea	against human captor, in jump-fights
Nose genitalia	—	P.verreauxi, L.macaco	males to females, in breeding season
Genital-mark	—	most lemuroidea?	males, females, at any time
Rub brachial gland on axillary gland	—	—	males, before tail or palmar marking
Palmar-mark	—	L.macaco	males any time, in stink-fights
Mark tail	—, purr, squeal	—	males, any time after rubbing glands and before tail-waving
Tail-wave	—, spat	—	males, any time esp. breeding season, in stink-fight

TABLE III-6

Lemur catta; VOCALIZATIONS

Vocalizations	Spp. with Homologous Signs	Occurrence
NOISES:		
Purr: faint click series..............	all lemuroidea	while mutual grooming, males marking tail
Repeated single clicks...............	—	staring at new object, mobbing ground predator
Click series........................	—	beginning of locomotion
Grunts.............................	—	in fast locomotion, mobbing
Explosive voiced grunt..............	—	rarely in intense mobbing
SPAT CALLS:	*P.verreauxi, M.murinus,*	
	L.macaco	
Infant squeaks......................	—	infant roughly handled, or losing contact with mother
Twitter............................	—	defensive greeting, usually male to female
Light yip..........................	—	defensive encounter
Deep spat..........................	—	defensive at high intensity, also while marking tail, when cuffing human being, juvenile when begging
Squeal.............................	—	while marking tail aggressively staring toward other male
Bark (yap).........................	—	mobbing predator, when neighboring troop barks
MEOW CALLS:		
Moan "mew",.......................	*L.macaco*	when separated from troop, answering meows or howls
Wail "me-ow",.....................	*L.macaco*	mew chorus esp. when troop changes direction
Howl..............................	*I.indri*	males: to change of light, before sleep, rarely to other howls, more frequent in breeding season
SHRIEK:.........................	*L.mongoz*	to hawks, or when seized by human being

second starts an untrue third above the first, like "singing" *Indri*.

Howls given in any circumstances are answered by a chorus of meows from the rest of the troop.

It seems likely that the howl functions as a form of self- and troop-advertisement. If normally given by an animal just before sleeping, it informs adjacent troops of the general sleeping location and so prevents encounters between troops in the sleeping trees. During the breeding season, it advertises the location of each male as well as of his troop and was, in fact, given by all males during the inter-troop meeting described above. It is typically given *after* mating and intermale disputes and thus again has some aspects of an attempt to regain contact with the troop in general.

The howl or song of *L.catta* thus has some features of a contact call: it is of the same form as the meow contact calls; an animal often starts with a few meows and breaks into howls; the howl is answered by meows from the troop; the howl is given (in breeding season) after fights and (in general) before going to sleep. On the other hand, it probably also functions to some extent as a territorial spacing call, which is heard by neighboring troops before settling to sleep. In the breeding season it is a spacing call when aggressive contact is likely between males of different troops.

In *Indri* the similar song is clearly territorial in function.

FRIENDLY RELATIONS: CONTACT, GROOMING, PLAY

Contact, grooming, and play are continuing social bonds. *L.catta* often sit together on a branch, clumping like social finches (Morris 1956, Sparks 1964), each animal with as much surface as possible in contact with the others. A *Lemur* clump is often spherical, with several heads emerging at scattered points of its surface and striped tails curled around. *Lemur* may also sit in a row, each curled up with one side touching the curve of the next animal. Occasionally *L.catta* form a locomotive as *Propithecus* do. Usually such a contact group contains only two animals, but it may gradually build up to six together.

If the observer is very close, he may hear animals purr with bouts of very faint clicks as they gain contact or as they groom.

Animals sitting in contact often groom each other. One animal presents the part it wishes groomed to the other. The second grasps the limb or clenches its fingers in a tuft of fur and licks or scrapes with the tooth-scraper. As with *Propithecus*, grooming is almost always mutual. Either both animals groom each other simultaneously, with both heads moving synchronously, or they alternate in bouts of 10 to 30 seconds each. All parts of the body are groomed, although the parts initially presented are often the underside of the forearm and the underside of the chin. Genital grooming occurs but is done from the side, the head of one animal buried between the other's thigh and body. Animals do not present for genital grooming, although females present to males for mounting.

Play, as usual, involves many varied postures. Animals may chase each other in circles, one attempting to jump on the other and wrestle. The jumper has an advantage, since it lands with all four hands and feet while the one beneath must use at least one hand to hold the branch. Jump-on-and-wrestle is the commonest *L.catta* game. Animals may also sit facing each other, like *Propithecus*, wrestling with hands and feet and mock-biting each other's ears. They may hang facing each other by hands or feet, kicking or grabbing each other with the free extremities. Juveniles on the ground occasionally have a mock jump-fight (see page 100).

TABLE III–7

Lemur catta; Friendly Interactions in Troops without Infants in 1963

Pair	Contact	Grooming	Play	Total Observed
Male-male............	–	3	–	3
Male-female...........	6	10	1	16
Female-female.........	1	10	–	11
Male-juvenile.........	1	2	1	4
Female-juvenile	11	12	–	23
Juvenile-juvenile.......	–	2	–	2
Not identified	9	18	19	46
Total............	28	57	21	105
Total Friendly inter- actions per hour..				1.9

One more contact or "greeting" gesture deserves mention. One animal which approaches another or passes another sitting on a branch often touches the tip of its nose briefly to the second animal's nose. The gesture does not seem particularly defensive since it occurs between any classes of animal with the members of any class taking the initiative. Rather, it seems to have the multiple functions of a general greeting.

The various kinds of friendly contact do not grade into each other as much as in *Propithecus*. Grooming is not so clearly allied to play, and animals often sit in contact for long periods without grooming.

Grooming and contact usually occur at the quieter times of day —at the beginning and end of the siesta and at the end of the sunning period. Play often occurs in the sunning spot in the early morning or in the leisurely evening progressions between the siesta and the various feeding spots. When the troop is feeding seriously in mid-morning or is in rapid progression, such contacts are unlikely.

Touching noses, contact, and grooming occur at all times of the year. Play is far more common out of the breeding season. During the breeding season adults behave sexually or agonistically, rather than playfully.

Table III–7 shows the observed friendly relations during 1963. It is difficult to interpret these, since observations were made on troops

TABLE III–8

Lemur catta, Troop 1; Friendly Interactions in 1964 [1,2]

Pairs	Touch Noses	Contact	Groom	Play	Total Observed	Total Expected
Male-male	1	2	11	–	14	13
Male-female.	31	15	46	–	92	60
Female-female.	10	6	30	–	46	48
Subadult male-male.	2	–	2	–	4	7
Subadult male-female	3	–	6	–	9	12
Subadult male-juvenile	–	2	3	–	5	9
Subadult female-male	2	–	3	–	5	7
Subadult-female-female.	–	1	2	–	3	12
Subadult-female-juvenile. . . .	2	1	5	–	8	9
Juvenile-male	4	6	8	–	18	47
Juvenile-female.	6	60	59	–	125	84
Juvenile-juvenile.	1	1	2	3	7	28
Subtotal.	62	94	177	3	336	336
not identified.	21	50	65	–	136	
Total.	83	144	242	3	472	

Total friendly interactions per hour: 3.1

[1] Includes the breeding season.
[2] Does not include interactions with the *L.macaco collaris* in troop.

of widely differing composition. Table III–8 shows friendly relations between the classes of animals in troop 1 in 1964. Figure 2 shows the absolute increase in number of friendly contacts in troop 1 during the early part of the breeding season; it fell off sharply during the week of actual mating. Figure 3 breaks down the proportions of male-male, male-female, and female-female contacts, showing that at the beginning and end of the season, male-male friendship was high, while females were in contact with each other or males with random frequency. During about 2½ weeks, including the week of mating,

WEEKS OBSERVED

FIG. 2.—*L.catta* Troop 1. Observed friendly interactions. Weekly averages (moving averages of successive ½ weekly periods) standardized for day length and hours of observation.

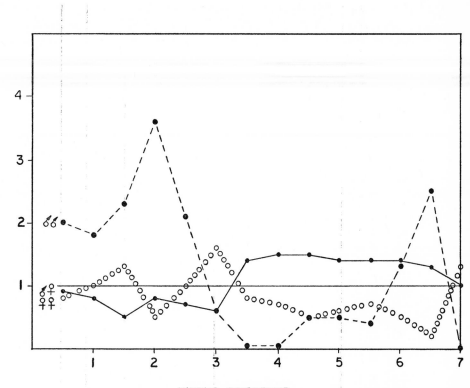

FIG. 3.—*L.catta* Troop 1. Proportions of friendly interactions between males and males, males and females, and females and females, standardized for number of animals of each sex. (The proportion of each kind of interaction, on the hypothesis of random interactions between all individuals, would be 1.)

TABLE III-9

Lemur catta, Troop 1; Friendly Interactions and Mating [1] between Individuals, [2] in 1964

	Male									Female						Subadult	
	AA	SC	PH	GN	BD	LN	RP	CT	RD	F[3]	VT	JE	D3	YT	BU	TL	IS
Females																	
AA																	
SC			1														
PH				1													
GN																	
BD				1													
LN																	
RP																	
CT				2	1	1											
RD				2				2									
Males																	
F[3]	2	2	1	5	2	2M	1	1									
VT	2	1	1	3	2												
JE		2		2		2M	1	2M									
D3		1		1		6M			1		3						
YT				m	m												
BU			M	m	m	M	2	M		1	1			9			
Subadults																	
TL	1	1		2	1		1					2	1	1	2		
IS				1	1						2	2	2				

[1] Mounting = m; Completed Mating = M.
[2] Individuals listed in order of dominance (Table III-11).
[3] F = male L. macaco collaris in troop 1.

females were more likely to sit with males and males were very rarely
with each other.

In general, the males of troop 1 who groomed each other were
numbers 4 and 5, the two most subordinate. They also tended to lag
behind the troop in the Drones' Club. During the breeding season,
these two males became aggressive, challenging each other as well as
their superiors, and ceased to groom each other. At the end of the
week of breeding, when all the males were battle scarred and limp,
males 2 and 3 sat together and groomed; males 1 and 2 both sat with
male 5 (Table III–9).

There seems to be no clear pattern in male-female friendly
relations, except that males 4 and 5 rarely sat with a female. The
three dominant males, however, seemed to associate with any female, although I saw the most dominant female only with the most
dominant male. There are not enough data to be sure of patterns in
female friendship.

AGONISTIC RELATIONS: DOMINANCE

GESTURES.—*L.catta* has a variety of agonistic gestures (Table
III–5). In the simplest situation one animal may stare at another; the
second animal narrows its eyes and looks away. One animal may
walk directly toward another; the second avoids the first or spats.
This walk, in dominant males, may become a real swagger, a stiff-
legged slightly rolling gait with head and tail in the air accompanied
by staring at the animal to be intimidated. On occasion, one male
may stalk another across 15 m, swaggering during the last 3 m.

In closer contact, one animal may displace another slightly by
pushing sharply with the end of its muzzle or by a butting gesture
with the top of the muzzle. This is a fairly low-intensity gesture,
used, for instance, by females toward juveniles which try to nurse
persistently or toward other females approaching to groom their
young infants. The nosed or butted animal rarely makes a spat call
but simply moves off or tries again to gain contact.

The commonest agonistic gesture is the cuff (Plate 10). In most
spats where animals come in contact one animal swats with one or
sometimes alternate hands toward the other's face. *L.catta* finger-
nails are long and pointed and quite capable of scratching. The full
cuff, though, ends with grabbing and tearing out a tuft of fur. This
happens most often during the breeding season. At other times the
cuff is usually just a harmless bat over the face or ear. The cuff
almost always accompanies a spat call from one of the pair.

Animals may simply feint instead of cuffing: they lunge forward with their shoulders but stay in the same place on the branch,

rather like fighting pottos (Bishop 1964). They may also lift one hand in a mock cuff. The feint is used both to warn off an inferior, a large approaching half-menace like a *Propithecus,* or a human being outside a cage.

Cuffing and feinting, of course, do little damage. Biting does real harm. *Lemur* have long upper canines whose points extend over the lower lip. In self-defense or in fights during the breeding season, *Lemur* make long downward slashes with these canines, usually raking one tooth like a nail through the opponent's skin. At the end of the breeding season, all five males had 5-cm slashes on their thighs, arms, or sides. None seemed disabled, but the one with an abdominal wound had narrowly escaped.

Spats are brief agonistic encounters defined as those situations when a spat call is given. They include almost all *L.catta* quarrels, except the rare supplanting, chasing, and jump-fights. The spat is a useful category, since it can be counted without seeing the animals.

Chasing and jump-fights I saw only during the breeding season. Males chased both other males and half-willing females. The chase was sometimes clearly aggressive, occuring between phases of a stink-fight or directed to challenging males between matings. Chases of females were longer, going on for perhaps 5 minutes and covering 200 m, and presumably sexual, at least on the male's part. Jump-fights were the most serious fights that I saw. These occurred on the ground. One or both animals rose on their hind legs with arms outspread, and they jumped or hopped around each other. An animal which could grab the other from above was thus in a position to give the downward slash of the canines. A jump-fight might also start in low trees, but if the animals grappled with each other there, they fell together to the ground and bounced apart. The longest jump-fight I saw was between a female and the *L.m.collaris* male who was part of troop 1. However, it is probable that the more serious injuries of the males were acquired during jump-fights.

VOCALIZATIONS (TABLE III–6).—Agonistic vocalizations are chiefly the spat calls. Animals may click or click-grunt loudly in threat to a predator and click-grunt before starting fast locomotion. Thus, the click-grunt is occasionally combined with a threatening stare toward another *L.catta* before locomotion toward the threatened animal. However, this is rare.

The spat calls are series of sharp squeaks, given with a grinning expression. They are derived from the infant discomfort call, a series of squeaks given, for instance, when the infant is groomed too hard by a troop member. The mother promptly reacts by collecting her infant and going off with it. Juveniles may use the deep spat in clamorous begging, either for contact denied them, as in weaning, or for food held by a female. Begging calls may be considered a transitional step between infant discomfort and adult social defensiveness.

Adults have a variety of spat calls. All are given in defensive agonistic situations. There are intermediates between them, but a few can be associated with particular situations.

The highest in pitch and softest, the "twitter," was given by males of troop 1 approaching a female just before the mating season. The male attempted to touch noses with the female, approaching slowly and hesitantly. The female almost always cuffed the male hard on the nose, rarely allowed herself to be groomed. Twittering did not precede sexual activity. It is not, however, clear that this call is purely agonistic (Andrew, personal communication).

The "yip" (Andrew 1963a) is somewhat louder and deeper in pitch. Andrew associated it with "friendly but defensive greeting." It was often given by a female cuffing a male who had approached her or by a caged animal reaching out to gratuitously claw a human being. It was also given by animals which crossed in front of a superior but were not directly threatened by it. It might be given by one animal relinquishing food to another. In general, the yip is a spat call of low intensity. It may either be given by the subordinate animal or by a dominant one which is closely approached and is cuffing the subordinate.

The deep spat is the same sound as the loud juvenile begging call. It is given by the defensive animal in a high-intensity spat, as when a superior animal ran or leaped toward an inferior.

One call which sounds related to the spat calls, although I am not certain, is the "squeal." This was a single, high-pitched note given only by males wrist-marking their tails at high intensity while staring toward another male preparatory to a tail-waving attack.

The spat calls grade into each other although typical twitters, yips, and deep spats are easily distinguished. They may all be considered as different levels of intensity of the same call.

I have described typical situations for the occurrence of each type of call in terms of dominance relations. However, I observed first that an animal which gave a deep spat was usually being chased by another and fleeing from it, while an animal yipping might stand its ground, cuff, and drive away the approaching animal. The dominance hierarchy is thus derived from watching results of spats, not vice versa. "Winning" a spat is defined as standing one's ground, "losing" as cringing or running away, irrespective of which animal gave the spat call.

SCENT-MARKING.—Both males and females genital-mark branches. They choose a small, near-vertical branch or stem, often smelling it first. Then they turn round, stand on their hands on a horizontal branch or the ground, hold on to the vertical branch with their feet as high as possible, and apply their genitalia to the branch, often rubbing up and down about 2 cm two or three times. In *L.macaco* the anal region is surrounded by glandular skin and the scrotum is furred; so the same gesture is more properly called ano-genital-marking (Andrew 1964*a*). In *L.catta*, the anus is surrounded by fur, and clearly only the genitals touch the branch: in females the vulva, in males the posterior surface of the bare, glandular scrotum. Frequently several animals in succession smell and mark the same spot.

Males also palmar-mark branches. They stand on their hind legs, grab a slender, vertical twig with one hand just above the other and jerk their shoulders horizontally from side to side. This simultaneously pulls their hands round the twig, leaving the palmar scent on it, and shakes the twig violently, so that it is possible to see one group of leaves in a tree jerking about and know that a male *Lemur* is marking beneath. Male juveniles begin to hold twigs at 6 months, but do not really mark. I once saw a subadult male (*ca.* 1½ years) hold two twigs, one in each hand, with the result that he could not fling himself properly from side to side around them, but oscillated back and forth.

By far the most dramatic of the male gestures is tail-marking and tail-waving. These involve the two main external glands of the male of *L.catta*, the antebrachial organ and the brachial gland (Pocock 1918). The brachial glands are large (2 cm in diameter) glands just above the axilla on the ventral surface. The single opening of each gland appears as a black protruding nipple high on the male's chest. Fatty secretion collects in the lumen of this gland and can be squeezed out the opening (Montagna 1962). The antebrachial organs are the black spur and gland on the inner forearms. The spur is true epidermal horn pierced by the openings of sweat glands. The glandular portion again gives a fatty secretion and is apparently an endocrine as well as exocrine organ (Montagna 1962).

The *L.catta* touches or rubs briefly the antebrachial organ to the brachial one. It is possible that the two secretions are thus mixed or that he simply picks up some of the brachial secretion on the spur. This anointing of one gland with the other may occur before palmar-marking with little apparent functional result and usually occurs before tail-marking.

The *L.catta* stands on his hind legs, tail drawn forward between his legs, and up between his forearms. He pulls his tail downward between three and ten times between his forearms. The spurs enter the fur deeply, and much of the tail must be in contact with the antebrachial gland.

At the same time the *Lemur* stares toward another animal. He draws his upper lip forward and down, so that it covers the points of his canines and protrudes somewhat below the lower jaw. This gives the front part of the lemur's muzzle, a square, houndlike appearance with two lowered flaring lips, but the lips are tense and do not droop like a hound's. This expression probably flares the nostrils. He may either squeal or purr while tail-marking. He then stands on all fours with tail arched over his back, its tip just above his head. He quivers the tail violently in the vertical plane, shaking its odor forward. Tail-waving is always directed toward another animal which may be right in front of him or more than 3 m away. The tail and body point toward the second animal, but the tail-waver may avert his head slightly.

The whole sequence of tail-marking and tail-waving seems to have a great aggressive content. The male toward whom the tail-waving is directed usually spats and runs. A female often spats and cuffs the tail-waver.

Anointing, tail-marking, and tail-waving are all characteristic of adult males. However, during one sequence of mating, both a 1½-year-old subadult, and a ½-year-old juvenile developed the full tail-marking and tail-waving pattern, although without anointing. It seems possible that the male scent does not develop until sexual maturity, for the female addressed by the subadult at first just ignored him, then sat in contact, with no start or spat to all his posturing.

A stink-fight is a long series of palmar-marking, tail-marking, and tail-waving directed by two males toward each other. The animals stand between 3 and 10 m apart. First one marks, then the other, with pauses between. Occasionally both tail-wave simultaneously, the two arched backs and tails reflecting each other like a heraldic design. The more aggressive male gradually moves forward, the other retreats, although they are often not close enough to supplant each other in one leap. The more aggressive one palmar-marks branches the other has marked. A stink-fight may go on from 10 minutes to an hour.

In 50 minutes of a stink-fight between males 4 and 5, male 4 palmar-marked twenty-nine times, male 5, fourteen times. Male 4 tail-waved seven times, male 5 only twice. Male 4 approached male 5, five times and supplanted him twelve times, while male 5 approached male 4 six times but supplanted him only once.

During pauses of a stink-fight the animals commonly make an exaggerated chewing motion, yawn, and self-groom. In the stink-fight above, chewing and yawning were only seen once each, but male 4 stopped four times to groom, and male 5, seven times. This chewing motion may have originated as a "rejection response" (Andrew 1963*a*), since it is seen in a juvenile who bites a rotten kily pod or in the animals nosing beetle colonies. It may still be a response to olfactory stimulus in the stink-fight, not just a displacement movement.

Scent-marking in *L.catta*, as in *Propithecus*, seems best classed as self-advertisement. It is often used in aggressive contact between troop members but not habitually in intertroop conflicts, which is quite different from *Propithecus*.

DAILY AND SEASONAL OCCURRENCE.—*Lemur* generally spat with each other during active periods of the day: the early morning progression, the mid-morning feeding period, and the late afternoon feeding and progression.

Figure 4 shows the increase in number of spats during the breeding season, rising to a high plateau about 4 weeks before the week of mating and falling off just before that week. Most stink-fights, and all jump-fights, occurred during the week of mating; so the fall in number of spats indicates the decrease of low-intensity aggression rather than of total aggression. Figure 5 shows the increase in scent-marking during the breeding season with its sharp peak of stink-fights in the week of mating.

DOMINANCE IN TROOP 1.—In *L.catta* troop 1, there was a clear dominance order among males, a fairly clear one among females, and females were dominant over males.

The concept of dominance has been loosely defined and defined in many conflicting ways (Gartlan, in press). Here, an animal is called dominant over another if it won consistently in spats and other aggressive encounters. In troop 1, during 1964, it was possible to rank all the animals in linear order, each animal being defeated by those above it and defeating those below it, although the exact rank of all females was not certain. The only aggressive encounters which did not fall into line are those with the *L.macaco* male in the troop (chapter IV) and several between males around a mating female. In this troop there seems to have been a linear dominance hierarchy, except during periods of actual mating.

Males and females had quite different general comportments (Table III–10). The males were usually on guard. The three more dominant animals carried themselves erect, stiffly, or even swaggered. They tended to stay in the front of the troop, among the females. They were constantly on the lookout for each other or subordinate males, whom they chased or otherwise threatened. The two more subordinate animals (and sometimes males 2 and 3) lagged behind the troop, carried head and tail lower, and were ready to run as another male approached. Male 3, especially, changed carriage and attitude abruptly, depending on whether he threatened or was threatened. He also frequently approached females and was frequently chased by male 1 (Table III–11). Thus, the male dominance relations expressed themselves not only in a large number of spats but in less easily quantified but even more frequent postures, glances, and avoidances.

Females, on the other hand, were far less "status-conscious." They might gratuitously chase each other or the males and would

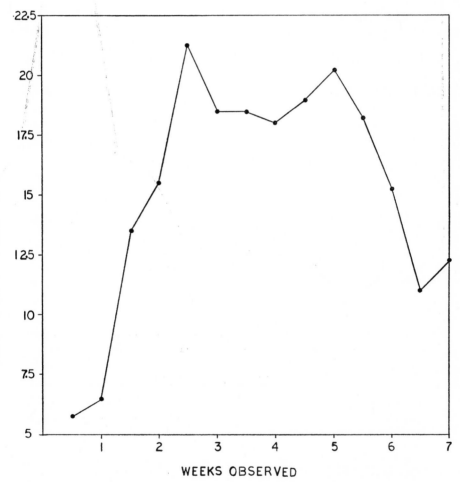

Fig. 4.—*L.catta* Troop 1. Total agonistic interactions. Weekly averages (moving averages of successive ½ weekly periods) standardized for day length and hours of observation.

SCENT-MARKINGS PER HOUR

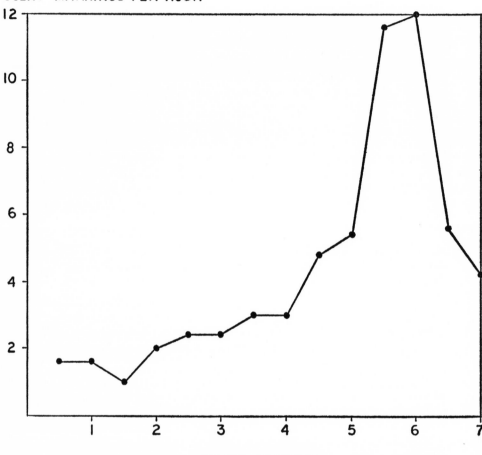

WEEKS OBSERVED

FIG. 5.—*L.catta* Troop 1. Observed scent-marking. Weekly averages (moving averages of successive ½ weekly periods) standardized for day length and hours of observation.

cuff any animal which came too close. However, not only did they spat less frequently, but they did not carry themselves in a particularly erect or cringing posture, nor did they keep an eye on dominant troop members and dodge their approach. The general dominance of females over males seems to rise out of the same attitudes: an insouciant female would cuff any animal, but a male was subordinate to any animal it could not bully. It may well be that if there were more data in Table III–11, the female hierarchy would show irregular-

TABLE III–10

Lemur catta, TROOP 1; AGONISTIC INTERACTIONS,[1,2] 1964 [3]

Pairs	Spats	Chases, Jump-Fights	Total Observed	Total Expected
Male-male................	179	32	211	25
Male-female..............	183	20	203	112
Female-female............	136	2	138	89
Subadult male-male	–	1	1	12
Subadult male-female......	3	–	3	22
Subadult male-juvenile.....	1	–	1	17
Subadult female-male......	6	–	6	12
Subadult female-female.....	9	–	9	22
Subadult female-juvenile....	3	–	3	17
Juvenile-male.............	5	6	11	86
Juvenile-female...........	31	2	33	156
Juvenile-juvenile..........	3		3	52
Subtotal	559	63	622	622
Not identified...........	421	12	433	
Total...............	980	75	1,055	

Total agonistic interactions observed per hour: 7.4

[1] Does not include tail-waving or stink-fights.
[2] Does not include interactions with the *L.macaco collaris.*
[3] Includes the breeding season.

ities, not a linear order. It may also be that there is a consistent order among the females, but that it is rarely expressed. One female was particularly aggressive, one so retiring that she had no spats with any animal in the troop, except the most aggressive female. Thus it is possible to see differences of general character among the females, even though they did not constantly behave as "dominant" or "subordinate" like the males.

Figure 6 shows the relative proportions of agonistic encounters among adults in troop 1 in 1964. Scent-marking and stink-fights are not included; spats, chases, and jump-fights are included. Male-male

TABLE III-11

L.catta, Troop 1; Spats Between Individuals [1]

VANQUISHED		Females									Males						Subadults	
	AA	SC	PH	GN	BD	LN	RP	CT	RD	F²[2]	VT	JE	D3	YT	BU	TL	IS
Females																	
AA										1							
SC																	
PH										1							
GN			1							1							
BD				2						1							
LN					1												
RP										1							
CT							2			1							
RD							2	2		1							
Males																	
F²[2]				1							1						
VT				2					3	5		3	7				
JE			1	2					1	1							
D3				2	2	3	2		4	5		34		8			
YT						1				2		2	7		10		
BU						1		1		3		5	5	3			
Subadults																	
TL				1	1												
IS	4			1	1					1							

[1] Excluding periods when mating was occurring.

[2] *L. macaco collaris* male.

PROPORTIONS OF AGONISTIC INTERACTIONS

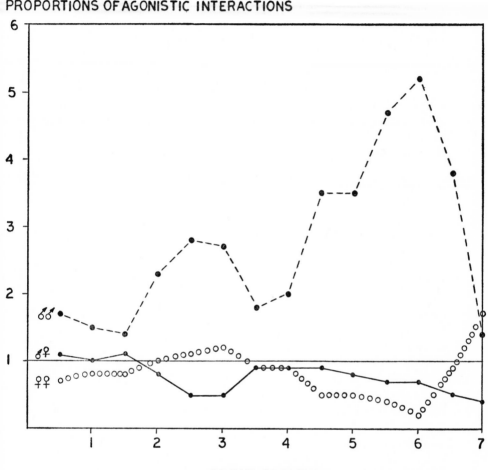

WEEKS OBSERVED

FIG. 6.—*L.catta* Troop 1. Proportions of agonistic interactions between males and males, males and females, and females and females, standardized for number of animals of each sex. (The proportion of each kind of interaction, on the hypothesis of random interactions between all individuals, would be 1.)

aggression reached a low 3 to 4 weeks before mating, and a high during the week of mating. Figure 5 shows there was even more male-male aggression expressed in stink-fights. Male-female aggression dropped between 3 and 4 weeks before mating, but rose again and decreased only gradually up to and after the week of mating. The males brought on many male-female agonistic encounters by approaching females, but the females always won; so this is a complex measure. Relative female-female aggression rose during the period 3 to 4 weeks before mating and fell steadily until the week of mating, after which it rose sharply. Total agonistic encounters (Figure 4), which would largely be made up by the rise in male-male aggression, were very high in the period 3½ weeks before and until mating. On the whole, it seems that the males become increasingly aggressive in the month before breeding and the females much less so.

It is significant that aggressive interactions in troop 1 seemed to take place between individuals, with no tendency to defend the underdog in a spat and with no central hierarchy of males who mutually defended each other, as in many baboon and rhesus troops (deVore and Hall 1965, Southwick *et al.* 1965). The two exceptions were females that came to the defense of a juvenile and dominant males after observing a male-male spat that often challenged the winner and put him in his place. These, however, do not approach the subtlety of baboon-macque relations in which animals may enlist each other's support, or simultaneously threaten another. Neither did I see "protected threat," that is, when one animal avoided the attack of a dominant by standing near another still more dominant (DeVore and Hall 1965, Kummer 1957). "Redirected aggression," however, was common, as in other vertebrates. Redirected aggression led to ever more intense male-male bickering; whereas in baboons, threatened males can let off steam by chasing a female, an *L.catta* male, gratuitously cuffed by either male or female, can only redirect toward a subordinate of his own sex.

SEXUAL RELATIONS AND BREEDING SEASON

FEMALE GENITALIA AND LENGTH OF SEASON.—Female *Lemur* have a visible estrus (Cowgill *et al.* 1962). In *L.catta* the genitalia swell from about 1.5 to 3 cm in length and develop a light, then a pink, center. I classified genitalia according to the size and color of the center, not on the color of the whole genitalia. The categories were:

Black
Small light: center smaller in diameter than black edge
Large light: center larger in diameter than black edge

> Small pink: flushed small center
> Large pink: flushed large center
> Bright pink: large, very flushed center. The vulval opening
> sometimes visible as a distinct hole.

Figure 7 is a diagram of the genitalia of individual females in troop 1. Counts of the troop make it clear that on March 17, at the beginning of study, all animals had either black or small light genitalia.

Four or perhaps five of the nine females went through a pink phase 3 to 4 weeks before the week of mating; then they faded before mating and flushed again. Two of these females were seen mating in the second period. It thus seems likely that there is an initial estrus or pseudoestrus period roughly 1 month earlier than the true breeding season, since *L.macaco*, at least, is polyestrous in captivity. Five of the troop reached the full bright pink condition. Four females were observed mating: one with large light genitalia, one with small pink, and two with bright pink. The one with light genitalia was a rather special case, having been pursued for some days by the *L.macaco* male, who may have prevented her from mating normally. The difference between small and large pink is not clear; possibly nulliparous females do not reach large pink condition.

In two instances in which females were observed both on the day of mating, and on the following day, their genitalia had faded considerably by the following day and were no longer pink two days following mating. From this, it seems reasonable to assume that the nine females of troop 1 all mated between April 16 (Aunt Agatha = AA) and the last observed mating April 27 (Grin = GN).

All six troop 2 females had large or small pink genitalia on April 23, when mating was also seen. Three females had faded to small light by April 29 and at least one more faded to large light by May 1. (One was pink a month before, on May 28, which may have been the earlier cycle.) Females with pink genitalia were seen in troop 3 on April 21, and in troop 4 on April 22. One female in troop 4 had the blazing pink center with the large vaginal opening usually seen just before and during mating. In short, the breeding season of troops observed in the Reserve seemed to last about 2 weeks, perhaps less for any one troop. Each female probably can mate for only 1 day a year.

There is no data for 1964 on possible later cycles. In 1963, however, *L.catta* seemed to be in breeding condition by April 6, when I left the Reserve. At that time at least one female of troop 1 was in the bright pink condition and there was a high level of aggression

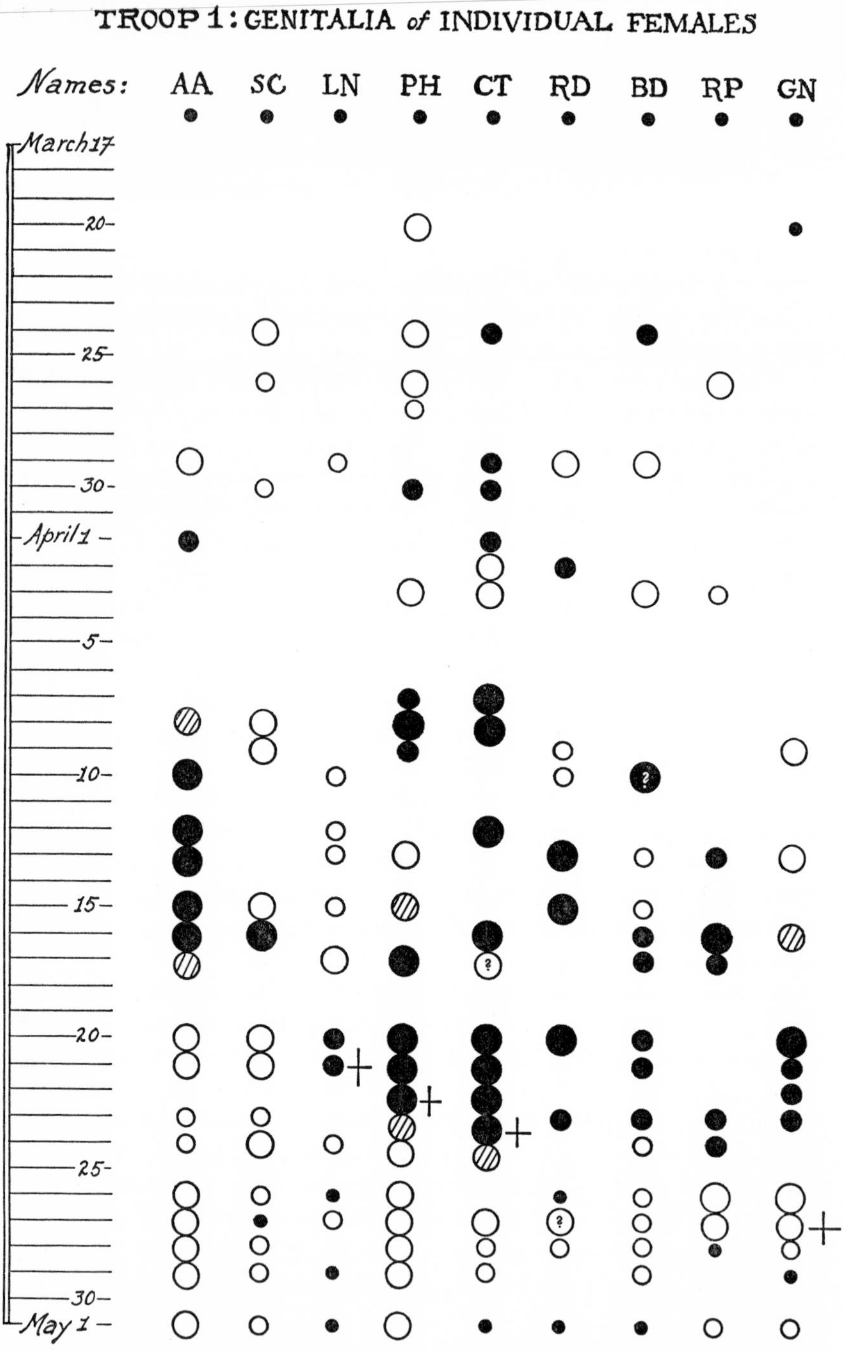

TROOP 1: GENITALIA of INDIVIDUAL FEMALES

FIG. 7.—Synchronized estrus in *L.catta* Troop 1. All female genitalia were quiescent, with no visible center, at the beginning of observations, though individuals could not be reliably identified. Unpigmented central area developed in March, flushing pink in a few females in late March, and in all females during estrus in April.

among the males. On May 11, when I returned, all females observed in the Reserve had black or very small light centered genitalia, and there was little activity of any sort in the troops. It thus seems unlikely that there is a significant post-seasonal estrus cycle, unless it is less than 1 month after the usual season.

Possible mechanisms and functions of this seasonal synchrony are discussed on pages 155 to 167.

PRE-MATING BEHAVIOR IN TROOP 1.—If mating began on April 16 and continued through April 27 in Figures 2 through 6, this is represented by weeks 4½ to 6½. All observed mating took place in week 6, although the moving average spreads the effect through weeks 5½ and 6½.

The absolute number of both friendly and agonistic encounters in the troop rose sharply from week 1. Friendly contacts reached a peak at week 4½, then fell, and in week 6 were largely replaced by true sexual behavior. Agonistic encounters reached a high plateau by week 2½ and began to fall in number in week 5, but they increased in intensity as spats gave way to stink-fights and jump-fights (Figs. 2, 4, 5). Figures 3 and 6 show the marked increase in relative male-male friendship and in male-male aggression during this period.

There is some indication of a cyclical effect corresponding to the early pinkening of female genitalia in weeks 1½ to 2½. The total friendly contacts rose sharply at this time and continued to be high, as though the changes in behavior began simultaneously with the first estrus, and continued until mating a month later. Male-male aggression reached a low peak at week 2½ (while male-female aggression decreased), then fell at week 3½, and rose to its highest at week 6, thus giving two cycles. In the friendly contacts there is a slight peak in male-female contact and also in male-male contact at week 2, at the expense of female-female contact. Scent-marking and tail-waving increased beginning with week 2, with a sharp peak in weeks 5½ and 6.

There were few qualitative changes in behavior until actual mating was observed. Males might groom females, but they did so in the ordinary manner, and tail-waving increased in intensity, perhaps, but not in form. No stink-fights or jump-fights were seen until week 6, the week of mating. (One prolonged stink-fight was observed between three males in August, 1963, when no females were breeding.)

On only two occasions were males seen to have erections while grooming females. The earliest of these, on April 10, was followed by a very weak attempt to clasp the female, who had small light genitalia and eventually mated (small pink) on April 21. The female cuffed the male severely; the male gave a deep spat and left. On the other occasion, April 15, the male did not attempt to mount.

COPULATION.—Four females of troop 1 with, respectively, small pink, bright pink, bright pink, and large light genitalia were seen mating on April 21, 22, 23, and 27. Three of the four females were seen mating in the morning between 9 and 12 A.M.; the other one and two of the first three also mated in the evening about 4 or 5 P.M. Mating possibly also takes place at night in the sleeping trees. Each female mated, or was sexually approached by males, during only 1 day, or if she also mated at night, a maximum of 36 hours.

All copulatory sequences began with at least two males approaching the female. The eventual mate either chased or stink-fought the others for between 5 and 25 minutes before mounting the female. The female presented, either standing with tail up and hind quarters raised, or else crouched flat against the branch in the "doormat" posture. The male then mounted the female, on either a horizontal or vertical branch, his arms clasped round her waist and feet gripping either the branch or her knees above the calf. Intromission rarely took place on first mounting, because after only 2 or 3 seconds, the male leaped away to chase one of the rival males. Three to twelve such mountings, with or without intromission and thrusting, took place, the male tearing away to chase his rivals up to twenty times between approaches to the female.

Twice juveniles, once a subadult female, and once (in troop 2) an adult (pink genitalia) female approached the pair during this period, and flung themselves on the male, cuffing and clawing. The male perforce dismounted to cuff or chase his attacker, but on two occasions when the male cuffed a juvenile, his female turned and cuffed him.

At last the male and female drew apart from the rest of the troop, because each time the male dismounted to chase, the female would move off by one branch (1 or 2 m). Copulation would then take place in periods of about 3 minutes each, the male thrusting rapidly and ending with ejaculation. There might be between two and four such periods. The periods of mating were separated by intervals when the male and female sat without touching each other, licking their own genitals, or else gently groomed each other's fur. After this the pair might leave together, presumably forming a consort pair for several hours, or they might return at once to the troop.

On one occasion, a female copulated with three different males in succession. The other females were seen with only one male at a time in either morning or afternoon.

Subordinate male number 5 accomplished three of six observed matings; so dominance status does not determine access to females, unless the dominant males had already mated (Table III–9). The dominance hierarchy was, in fact, invalid for agonistic interactions during periods of mating (Table III–12).

It would be quite easy to miss observing a mating if the consort pair had already left the troop but almost impossible to miss the chasing among males which preceded mating. Thus five of nine females were not seen mating.

MOTHER-INFANT RELATIONS

Naturally, infant *L.catta* are almost all born at once, after a gestation period of about 4½ months. In 1963, in troop 1, one infant was born August 20 or 21, the next three between August 22 and 27, and the remaining three sometime thereafter. In troop 2, at least two infants were born between August 20 and 22, and in troop 3, infants had appeared at least by September 3. In the troop at Bevala, 30 km away over sisal fields, infants were born sometime in the week before September 1. All these infants were about the same size.

The infant *L.catta* is about 10 cm long with large head and spindly limbs. It is rather brownish, but otherwise has adult coloring, down to a striped black-and-white tail as long as head and body, but so thin that it looks like a dashed line drawn in soft pencil on a piece of white paper. The infant clings longitudinally under its mother's chest, not transversely on the abdomen like *L.macaco* or *Propithecus*.

Infant *L.catta* are far more precocious than (captive) newborn *L.macaco*. As young as 3 days, when an *L.macaco* baby just alternates between nipple and crotch, infant *L.catta* move about actively on their mother's body, even onto her back or onto another grooming female. Petter-Rousseaux (1962) writes of a (captive) *L.catta* mother leaving her baby alone briefly at 14 days, although she was recalled by a distress cry. At 1½ months this baby was nibbling solid food. In February, 1963, I watched a half-grown juvenile alternately suckling and eating kily leaves. Some females still had black hairless breasts in early March, 1964, and juveniles twice put their mouths to the nipple for several minutes, although they seemed to use it more as a pacifier than for suckling. Juveniles still sat with the females frequently throughout the breeding season (Table III–8). There is not, at this season, clear preference for particular females.

Females and juveniles of the troop are quite fascinated with the newborns. Many attempt to groom the infant, but the mothers usually allow only other mothers to approach (Table III–13). Then they

TABLE III-12

L. catta, Troop 1; Spats during Periods of Observed Mating

VANQUISHED	VICTOR — Male									VICTOR — Female						VICTOR — Subadult	
	AA	SC	PH	GN	BD	LN	RP	CT	RD	F[1]	VT	JE	D3	YT	BU	TL	IS
Females																	
AA																	
SC																	
PH																	
GN																	
BD																	
LN																	
RP																	
CT																	
RD																	
Males																	
F[1]				3			2					1			2		
VT							1			1				2	1		
JE																	
D3							1	3		4		2			5		
YT										2	5	1	1		4		
BU							1	2		1		7	1	1	4		1
Subadults																	
TL																	
IS														1			

[1] F: male *L. macaco collaris*.

reciprocally groom each other's infants. Thus, after the birth of infants, I saw mothers groom their infants eleven times, *other* mothers groom infants thirteen times, males not at all, and all other females and juveniles groom only fourteen times.

TABLE III-13

Lemur catta; Friendly Interactions in Troops with Infants

Pair	Contact	Groom	Play	Total Observed
Male-male.............	–	–	–	0
Male-female..........	1	1	–	2
Female-female........	–	–	1	1
Male-juvenile.........	–	1	1	2
Female-juvenile.......	–	–	1	1
Juvenile-juvenile......	–	–	–	0
Male-Mother.........	1	–	–	1
Male-infant..........	–	–	–	0
Female-mother........	2	–	–	2
Female-infant.........	2	5	–	7
Juvenile-mother.......	4	–	–	4
Juvenile-infant........	–	7	–	7
Mother-own infant.....	–	11	–	11
Mother-other infant....	1	12	–	13
Not identified.........	–	–	8	8
Total.............	11	37	11	59

Total friendly interactions per hour: 3.3

August 23, 1963. Two mothers of troop 2 sat in low lianas, each with an infant no more than 3 days old; Mary's was less active and with less fur, especially on the tail. Elizabeth approached Mary, groomed Mary's baby, and sat in contact with Mary. Elizabeth's baby scrambled around onto her back. Mary groomed Elizabeth's baby and then her own, while Elizabeth groomed her own. They sat a moment; a juvenile approached and sat in contact with the two mothers. The juvenile groomed the tail of Mary's infant, grabbing it in both hands and tooth-combing the wrong way of the fur, which parted to the roots. Mary cuffed the juvenile, who left. Mary and Elizabeth groomed each other's infants. Elizabeth's baby climbed half-way onto Mary's neck. Elizabeth then groomed her own baby. A female approached and sat in contact with Elizabeth, who cuffed her. She left, but a second female attempted to groom Mary's infant, so Mary cuffed the second female, who also left. Mary and Elizabeth groomed their respective infants. Mary's baby clung passively, but Elizabeth's crawled about, even on Elizabeth's back. The infant then transferred itself from Elizabeth's stomach to Mary's back. Mary rose and started to walk off with one baby clinging above and one beneath. Elizabeth watched calmly, but then she dropped her nose to her stomach, started, and walked after Mary. She pawed her baby off Mary's back and onto her own belly fur and then vigorously groomed Mary's baby. Both mothers somehow discouraged a third female from approaching. They separated, groomed, and allowed their respective young to suckle.

Thus, for the newborn *Propithecus* the world must be full of brothers and uncles, and when it grows up, it becomes part of a homogeneous group of equal companions. For the newborn *L.catta*, the world is full of mothers and a fringe of less privileged, near-mother females and juveniles. For the grown female *L.catta*, the troop continues to be a body of females and juveniles with males who are sometimes exciting or obstreperous, but the grown male *L.catta* joins a world where male status is the key to life, females are frightening, desired, or comforting and juveniles barely exist.

Chapter IV

Aspects of Behavior of *Lemur macaco*

DESCRIPTION AND RANGE

In the forest at Berenty, one male *Lemur macaco collaris* was a constant member of *Lemur catta* troop 1 (Plates 12, 13). This animal had been captured as an adult in the forest near Fort Dauphin and taken to the Mandrary Valley, where he escaped and joined the wild *L.catta* of the region. Other *L.macaco* subspecies have been studied both in the wild (Petter 1962*a*) and in captivity (Andrew 1963*a*, 1964*a*; Jolly 1964*a,b*). Therefore, this animal is particularly interesting because we can guess at some of his social relations in a normal troop of *L.macaco* and see how these became changed as he adapted, more or less, to life with the *L.catta*.

The *L.m.collaris* was about the same size as the *L.catta* and very much the same shape. However, instead of having their gray and black-and-white coloring, his body tone was rich chestnut with a darker tail. His face was black with white clown eyebrows and an orange beard with emerged in two bushy tufts on each lower cheek. Although it has not been established that *Lemur* can see color (Bierens de Haan and Frima 1930), it is highly likely that the color differences serve as recognition marks between species. The *L.m.collaris* therefore had a completely different set of recognition marks from the *L.catta*.

L.m.collaris also normally live in a different ecological area. In chapter I, I described how climax dry southern scrub can change over only 20 km to climax eastern rain forest in the mountains. The range of *L.m. collaris* is confined to the wet eastern mountains, the range of *L.catta* to dry transition forest, gallery forest, and western zone woodland.

The species *L.macaco* is one of the most widespread and diverse of lemur species. There are at least eight subspecies. They are distributed all around the periphery of the island, from Fort Dauphin in the southeastern corner, northward through the eastern forest domain, in the Sambirano zone in the northwest, and throughout the western forests. The only area of forest where *L.macaco* does not exist is the dry southern domain, where it is replaced by *L.catta*. *L.catta* overlaps extensively with *Lemur macaco rufus* in the west, but on the east its range touches the range of *L.m.collaris* without overlapping. (This division is true of most of the plants and animals of southeastern Madagascar which are confined either to the wet east or to the dry south.) Thus, while *L.m.collaris* very likely intergrades with other forms of *L.macaco* to the north, that is, in the eastern forest, we see a very sharp boundary of both anatomy and ecology between *L.m.collaris* and its neighbor on the west, *L.catta*.

ECOLOGY

L.m.collaris lives in forests where the rainfall approaches or even exceeds 300 cm a year. In the Mandrary Valley Forest, the rainfall was about 50 cm a year. The first difference in ecology then is that *L.m.collaris* would normally be able to drink from leaves or hollows in trees at any time. The *L.m.collaris* in the Mandrary Forest drank no more often than the *L.catta* it was with. Second, its normal food plants should have been entirely different, since there are almost no species which live in both forests, although a few kily trees can be found on the east coast. Some of the plants which *L.m.collaris* eats in the wild, according to native guides, are *Harungana madagascariensis, Tristemma virusanum,* and *Trema orientalis.* These are all lush green plants of which the animal eats either leaves or fruit. In the reserve, the *L.m.collaris* ate the same plants as the *L.catta*. I never noticed him either refusing a plant which they ate or seeking out a particular, different kind.

There are also differences in choice of branch height between normal *L.macaco* and *L.catta*. *L.catta* is probably more terrestrial in habit than other *Lemur*. In the wet eastern forest there is very dense foliage from ground level to about 21 m which is the top of the thick layer of forest. Therefore, there are branches to walk on over a much more continuous vertical height than in the forest at Berenty, where the animals tended to move from 6 to 21 m feet or else on the ground. Again, the *L.m.collaris* which was with the *L.catta* troop took the same routes that they did, both walked and fed on the ground, and did not prefer arboreal routes.

This lone *L.m.collaris* was, to all appearances, a healthy animal with springy step and glossy coat. It seems, therefore, that the life and diet of the *L.catta* were at least adequate for him and the ecological differences between *L.catta* and *L.m.collaris* are not sharply determined by physical factors of climate and food. It is possible that the two species are incompatible; that is, that *L.catta* and *L.m.collaris* enter into direct competition. However, in the western forest *L.catta* coexists with *L.m.rufus* so there may also be that there is more ecological difference than appeared.

INDIVIDUAL BEHAVIOR

The individual behavior of *L.macaco* is very similar to *L.catta*. The general activity is about the same. They sleep in the same positions and move and jump the same way; both species sun in the morning; both species groom themselves with their teeth and scratch with their toilet claws. They feed similarly, using mouth and hands only to draw in branches or occasionally to hold small food but with the mouth doing all fine manipulation. These activities have been well described for *Lemur macaco macaco* (Petter 1962a).

COMMUNICATION

Many gestures of *L.m.collaris* are essentially the same as those of *L.catta*. Like *L.catta*, the animal purses its mouth to wail during the contact call, it grins in fright or social submission, it carries head and tail erect when confident, and it hunches over with head and tail drooping when submissive. Tactile gestures of contact, grooming, and play are also similar.

However, the vocal communication is quite different (See Table V–1 and discussion pages 136–42. *L.macaco* vocalizations have been listed by Andrew (1963a). First is the click-grunt and grunt. *L.macaco* click-grunts are much more blurred than the clicks of *L.catta*. Even among the *L.macaco* subspecies, *L.m.collaris* has a particularly juicy voice with the clicks blurred together and a good deal of tonal vocalization. These grunts are given at the same time as *L.catta* clicks and grunts, when the troop is contemplating locomotion or when it sees a strange or unusual object; they seem to have the same function as contact and arousal calls. *L.macaco* grunts are given much more frequently in contact situations, as in greeting between animals when they meet, and in this usage correspond to the *L.catta* wails of contact, the meow calls.

The grunt type of call has a number of variations, named by Andrew (1963a) the querulous grunt, the bark, and the sharp call. Barks are loud calls given in much the same situation as the *L.catta* yap, that is, when mobbing a strange creature such as a dog, cat, or human being. Querulous grunts and sharp calls correspond, respectively, to the light spat and deep spat; that is, they are given in contact with a superior who is handling the animal roughly or else while cuffing another animal. The single *L.m.collaris* frequently gave locomotor and contact grunts at high intensity but rarely gave the other versions of the grunt. He did not take much part in mobbing me or other animals, and since he had a fairly dominant position in the troop, rarely gave the calls of a subordinate animal.

His most noticeable use of the grunt was to answer *L.catta* meow choruses. *L.macaco* has a wailing call which is homologous to the *L.catta* meow. The *L.catta* meow is a frequent contact call between members of the troop, however, while the *L.macaco* wail is given only by animals which are separated and lost or by extremely lonesome subordinates in a captive group. When the *L.catta* meowed, the *L.m.collaris* answered with very rich, juicy contact grunts, presumably responding as to the very high-intensity contact call of his own kind.

I did not hear this *L.m.collaris* give either the moan or wail. Neither did he make squeaks or shrieks in response to hawks.

The vocalization which Andrew calls a cough is particularly interesting. In *L.macaco* it is a sound which develops from short, quick grunts like the bark, but is much prolonged. It has a great deal of both masking noise and fundamental tone. It is given in a large number of situations, and its function may differ among the subspecies of *L.macaco*. When a group of *L.m.macaco* is disturbed, when one animal is isolated, or when two groups meet, they make long, rattling coughs, and other animals of nearby troops often cough at the same time (Petter 1962a). Petter also records a variant of this sound given in chorus at dusk as the animals were settling to sleep, which may also be true for *L.m.collaris* troops (Foiret, personal communication). The cough appears in many situations of great excitement or high stimulus contrast in the laboratory. (Andrew 1963a). The only time I heard the Berenty *L.m.collaris* give this cough was during a bout of howling by three males of the *L.catta* troop at the height of the breeding season. It was dusk; the troop was about to settle. The *L.catta* began to howl in a chorus almost like *Indri*, each animal starting one note later and an untrue third higher than the one before it, so that the howls overlapped. Abruptly, the *L.macaco* began to bark with the noise rapidly rising to a rattling cough and joining the chorus of wailing song.

In olfactory communication *L.macaco* again differs sharply from *L.catta*. *L.macaco* does not have the arm glands of *L.catta*, does not mark its tail, and has a furred scrotum rather than *L.catta*'s bare glandular one. However, *L.macaco* has a large, naked glandular area around the anus running down toward the genitalia (Montagna, Yasuda, and Ellis 1961). *L.macaco* marks by pressing its anogenital region against a branch and rubbing. It also presses its forehead against a branch, that is, the white eyebrows above the eyes, and rubs back and forth. It may urinate and defecate on branches. Petter (1962*a*) reports seeing animals rub their foreheads in their excrement and carry this as a mark. *L.macaco,* both male and female, often mark each other, rubbing the anogenital area on the flank of the other animal. Petter (1962*a*) reports a sharp increase of scent-marking during the April and May breeding season.

The *L.m.collaris* with the *L.catta* troop scent-marked, although far more rarely than an ordinary *L.macaco* male, usually by pressing his anogenital region against a branch. He often attempted to rub it against another animal, always a female *L.catta,* but was never allowed to do so, since the *L.catta* would move away abruptly. He did not mark by rubbing his forehead in excrement, although he occasionally rubbed it against a branch, especially where a *L.catta* had just been marking. He took no part in the stink-fights of the males, although he did approach the females and attempt to mark them. To the human nose, *L.catta* has almost no scent at all, but *L.m.collaris* smelled very strongly, particularly in the breeding season. At 3 or 4 m away it was perfectly obvious that he was there. The *L.catta* presumably found his scent distasteful, although since they found him generally distasteful, I have no behavioral indications to be quite sure that it was the scent that upset them.

SOCIAL BEHAVIOR

A normal *L.macaco* troop seems to be rather smaller than troop 1 of *L.catta,* averaging about nine members (Table IV–1). Several *L.m.macaco* troops may, however, sleep together (Petter 1962*a*). The *L.m.collaris* was very much a member of troop 1. He was never seen alone, but always with the troop. He moved with them in the position of a dominant male, that is, well forward among the females.

His relation with the males of the troop was that of the dominant animal by 1964, although in 1963 this had been less clear. That is, any male of the troop would step aside for him including the *L.catta* male 1. In one spat this *L.catta* did rout the *L.macaco,* but although his position may have been equivocal with respect to male 1, he consistently ranked ahead of the other four males (Table IV–2). His agonistic encounters with the males, however, were either spats in which one animal cuffed the other or, more often, situations

TABLE IV–1

Lemur macaco; COMPOSITION OF TROOPS

Date	Group	Male	Female	Juvenile	Total
May, 1956 [1]	1	7	3	–	10
	2	7	3	–	10
	3	2	2	–	4
	5	5	4	–	9
	6	5	5	–	10
	Total	26	17		43
Nov., 1956 [1]	1	5	3	2	10
	2	6	4	3	13
	3	4	2	–	6
	4	5	2 or 3	2	9–10
	4 bis..............	5	5	2	12
	5	4	4	1	9
	6	5	5	1	11
	7	3 or 4	3	–	6 or 7
	7 bis..............	5	3	1	9
	8	5	4	1	10
	Total	47 or 48	35 or 36	13	95 or 97
Apr., 1957 [1]	1	6	4	–	10
	2	6	7	–	13
	3	4	2	–	6
	4	5	4	–	9
	4 bis..............	8	7	–	15
	7 bis..............	5	3	–	8
	Total	34	27		61
Dec., 1962 [2]	A.................	4 or 5	2	2	8 or 9
	B.................	4	2	1	7
	C.................	2	2	3	7
	D.................	4	4	1	9
	E.................	5	4	2	11
	Total	19 or 20	14	9	42 or 43
Apr., 1962 [3]	A.................	3	2	1	6
	B.................	3	2	1	6
	Total	6	4	2	12

[1] From Petter 1962a, for *L.macaco macaco* on Nosy-bé and Nosy-Komba.
[2] Personal observation, *L.macaco macaco*, Nosy-bé.
[3] Boggess and Smith, personal communication, *L.macaco rufus*, Lambomakandro.

TABLE IV-2

L.macaco collaris MALE; INTERACTIONS WITH TROOP 1 *L.catta* 1964, EXCLUDING PERIODS OF OBSERVED *L.catta* MATING

	L.catta Groomed	*L.catta* Approached	*L.catta* Accepted Contact	*L.macaco* Approached	Spat, *L.catta* Vanquished	Spat, *L.catta* Victor
FEMALES:						
AA	–	–	–	2	–	1
SC	–	–	2	2	1	–
PH	–	–	1	4	–	1
GN	–	–	3	24	1	1
BD	–	–	–	2	1	–
LN	–	–	–	7	1	–
RP	–	–	–	2	1	–
CT	–	–	1	2	1	–
RD	–	1	1	2	–	–
not ident.	–	1	1	60	15	–
Total	–	1	8	107	20	2
MALES:						
VT	–	–	–	–	5	1
JE	–	–	–	–	1	–
D3	–	–	–	–	5	–
YT	–	–	–	–	2	–
BU	–	–	–	–	3	–
not ident.	–	–	–	–	5	–
Total	–	–	–	–	21	1
SUBADULT MALE						
TL	–	–	–	–	1	1
JUVENILES	3	–	6	6	–	–

where the L.catta simply ran spatting from the L.m.collaris. He did not take part in the L.catta scent-marking or stink-fights. He did not seem to take part in the usual interplay of glances and posture by which each male asserted its status to every other male within sight. It seemed a much more adventitious thing that when the two animals came close the L.catta gave way or ran away. I did not count as agonistic encounters situations where the males simply avoided meeting the L.m.collaris.

The L.m.collaris continually attempted to approach females of the troop, to sit next to them, and to groom them. The females' almost universal reaction was to get up and move away. At times, a female would cuff him, (Plate 14), or even cuff and spat, but at other times she would run away making spat noises. Thus, the L.m.collaris is the only male of the troop who could be said to have won an agonistic encounter with a female. Of all the females, only Aunt Agatha, the most aggressive, actually picked a fight with him and won (Table IV–2).

The L.m.collaris made sexual advances to the females during the breeding season; that is, he would chase females who were in estrus and attempt to sit next to them, although I never saw him attempt to mount, largely because the females would cuff him violently and run away as he approached. He did, however, initiate some of the sexual excitement in the troop because of his constant attention to the females (Table IV–3). He often provoked chases in which other males would then join. The first jump-fight that I saw was between the L.m.collaris and female "Lonelyhearts," who was also the first I saw to mate.

The L.m. collaris chased Lonelyhearts off and on from 8 to 10:30 A.M. on the morning of April 21. Twice during the morning, Lonelyhearts turned on him. She stood on her hind legs and danced around him, batting and cuffing with her hands and pulling out tufts of fur. He retaliated by also standing on his hind legs and pulling tufts of her fur. I did not see either bite the other, but the palm of her hand was slashed open. Lonelyhearts, having beaten off the L.m.collaris at 10:45 A.M., lay panting at the foot of a liana. She was approached by three males in turn. She eventually mated with male 3. Both returned to the troop, where male 3 harried other troop males. He retreated spatting when the L.m.collaris approached. The L.m.collaris chased, then grappled with, Lonelyhearts. Both fell off a liana for 3 m, still clawing each other. The L.m.collaris desisted temporarily, sitting alone while male 1 and male 4 had a long stink-fight on the ground. Male 1 then mated with Lonelyhearts, tearing away at 30-second intervals to chase male 4. After they had completed mating, over about 5 minutes, the L.m.collaris chased both of the pair away. Finally, male 5 tail-waved and chased the L.m.collaris and mated with Lonelyhearts, dismounting once to chase the L.m.collaris and once to fight a subadult female who had flung herself on the pair. Male 5 and Lonelyhearts then drew apart from the troop about 12:30, probably returning only 3 hours later as a consort pair.

TABLE IV-3

L.macaco collaris; INTERACTIONS WITH TROOP 1 *L.catta* DURING PERIODS OF OBSERVED *L.catta* MATING

	L.catta Groomed	*L.catta* Approached	*L.catta* Accepted Contact	*L.macaco* Approached	Spat	*L.macaco* Chased *L.catta*	*L.catta* Chased *L.macaco*	Jump-Fight
FEMALES:								
GN............	–	–	2	7	5	3	–	–
LN............	–	–	–	–	3	7	–	4
non-mating female....	–	–	–	4	–	3	–	–
TOTAL........	–	–	2	11	8	13	–	4
MALES:								
VT............	–	–	–	–	1	1	–	–
JE............	–	–	–	–	4	–	–	–
D3............	–	–	–	–	2	–	–	–
BU............	–	–	–	–	1	–	2	–
not ident........	–	–	–	–	1	–	–	–
TOTAL........	–	–	–	–	9	1	2	–
JUVENILES:	1	2	–	–	–	–	–	–

One female, Grin, became his particular favorite. During the breeding season I saw him attempt to sit with her twenty-four times and actually accepted by her—that is allowed to sit in contact—five times. He chased her particularly during the final days of the breeding season. On April 27, Grin was approached by several males of the troop at about 7:30 A.M. Male 4 mounted with thrusts, but the *L.m.collaris* chased both male and female apart before mating could be completed. During that day several males approached her, including males 2, 3, and 4, who vied with each other, and male 5, who actually mounted. However, in each instance the *L.m.collaris* chased away the other males. It is even possible that he prevented Grin from mating this season. He certainly disrupted most of the attempts to mate with her, although very likely some male actually did succeed.

The *L.m.collaris'* relations with the juveniles of the troop were much more cordial (Table IV–2). He rarely approached them, usually only after being rebuffed by several females, but when he did they would occasionally let him sit next to them or even groom him themselves. Twice I saw juveniles approach and sit with him. It will be very interesting to see in another year or so whether these juveniles, as adults, will accept the *L.m.collaris* more readily than their parents do.

From what is known of *L.macaco*'s social life in the wild, many aspects might be the same in normal *L.macaco* troops. *L.m.macaco* keeps to very limited territories (Petter 1962a); so once having joined a particular *L.catta* troop, as this *L.macaco* has, presumably he would attempt to stay in the same geographical area with the troop rather than wandering over the forest. The *L.macaco* troops are smaller, but he had very little choice in this respect. Petter (1962a) points out the preponderance of males in troops of *L.m.macaco*, like those of *P.v.verreauxi* (Table IV–1).

Male-male interactions of *L.catta* are largely by scent communication, that is, palmar-marking and tail-marking. The *L.m.collaris* was completely unimpressed by the *L.catta*'s tail-waving threats. He owed his dominant position among the males partly to this. The sexual behavior of *L.macaco* is not very well known, although Petter (1962a) has studied it to some extent and shows that scent-marking of one animal by another plays a fundamental role in the sexual behavior. The *L.m.collaris,* of course, never accomplished this. However, the breeding season of *L.m.macaco* at the northern end of Madagascar is the same as that of *L.catta* in southern Madagascar; so possibly all *Lemur* reach the sexual condition at about the same time.

Laboratory troops of adult *L.m.rufus* and *L.m.fulvus* indicate that the difference in temperament between male and female *L.macaco* may be roughly the same as in *L.catta;* that is, the males all are greatly concerned with dominance and spend a large proportion of their time in agonistic encounters with other males, while the females have a much less marked hierarchy (Jolly, in press). This shows particularly in their relations to human beings. Wild-captured adult males are difficult or impossible to tame and run from a human being or attack him as they would a superior or a threatening animal, whereas females are extremely easy to tame, seeming to have no fear, and therefore readily accept a human companion. The *L.m.collaris'* lack of awe for the *L.catta* females may be due to another breakdown in communication between the species, like the one involved in his dominance over the males.

In conclusion, this *L.macaco* had a definite place in *L.catta* troop 1, even though an uneasy place, largely due to his own efforts to approach members of the troop. He seems also to have no difficulty in coping with the ecology of *L.catta* in the gallery forest.

Chapter V

Displays

FORM, FUNCTION, AND MOTIVATION

There are several ways to classify displays. The first, now accepted as fundamental, is description of *form*. The description of form, however, leads to more speculative classifications by either *motivation* of the displaying animal or by *social function* in the group.

We assume that the three types of classification bear some relation to each other. A single element of display is generally assigned a single motivation and a single social function. A second, quite different assumption is that displays of similar form are likely to have related motivations, even if they are now distinct (Darwin 1872, Andrew 1963*a*, 1964*a*, Altmann, 1962). This second assumption underlies modern phyletic comparisons of displays: for instance, the attempt to trace a prosimian protective grin through the gradual changes of motivation that led to the baboon-macaque "fear grin" in social contexts and finally to the human smiles (Andrew 1963*b*). It also justifies some grouping of displays and motivations within a species; as, for instance, I relate *L.catta*'s howl to the meow group contact calls, because it is so clearly similar in form.

The first assumption, the equating of motivation and social function, underlies much classic ethological theory and many descriptions of primate behavior. Often, we do not bother to distinguish between motivation and function: phrases like "confident threat" or "alarm calls" imply both.

This does not work so simply with lemurs, because we know too much about their displays. Andrew (1962*b*, 1964*b*) has extensively analyzed the stimuli which evoke prosimian vocalization and facial expression in the laboratory among animals exposed to a large number of abnormal situations. It is clear that stimuli of many different kinds may evoke a single display, so that we cannot assign the motivation of the display to a particular situation, or even to a single emotional "drive." Andrew describes these displays as resulting from "stimulus contrast." When a stimulus of any sort rouses the animal slightly, it will give one kind of call; when it is roused more, it gives a different set of calls. The "contrast" of the stimulus can be increased by previous expectation. Chicks, for instance, can be conditioned to vocalize to the noise of an approaching food hopper. Andrew points out that human laughter is a response to moderate contrast in many situations, while we cry to stronger contrast in the same situations in a close analogy to the chicks' twitter and peep. We smile or laugh to see the face of a friend after separation but weep in helpless joy over a son returned alive from the wars.

Perhaps the clearest demonstration of stimulus contrast is bouncing a baby. At the first bounce or so the child's face is blank. As the baby begins to expect bouncing, the stimulus contrast is raised; it first smiles, then laughs compulsively. If the bouncing goes on too long, the baby can bear no more stimulation. Its open mouth turns down at the corners, its lower lip pushes forward and up, the laughter grades into a hiccupping sob and finally full-blown howl.

The motivation of some displays can be ascribed to "drives" like aggression or sex; others can be ascribed to stimulus contrast. In each case, though, the presumed individual motivation is a far more generalized concept than the functions we observe in the wild. There, because the display occurs in specific situations, it evokes precise and appropriate responses.

We must look at the context of a display element to understand its social function. This context may be other signals given by the same animal (Marler 1965). It may be a social background, such as the rank of the displaying animal. It may even be an environmental situation. If we take account of the context, the meaning or function of a display is usually perfectly clear both to the observer and to watching animals.

Context is just as important to the most sophisticated of primates. Darwin (1872) showed that human facial expressions are often very difficult to interpret, unless we know what situation has given rise to them. When we have a normal amount of information about the situation, however, we find each other's expressions full of precise and subtle meaning.

One example might be the shriek of *L.catta*. In the laboratory, the shriek seems motivated by the simple emotion of terror—when an animal is seized with a pair of heavy leather gloves. In the wild, *L.catta* shrieks almost exclusively at flying hawks. Other troop members who may not have seen the hawk all join in the cry and at once leap downward in the trees under cover. (We do not know if the hawk stimulus is innate or learned.) The shriek, though motivated by terror, functions as an air-raid alarm.

A more complex example is *L.macaco*'s cough. This appears in captivity during excited locomotion, when frustrated by a closed door, when the sounds of approaching breakfast are heard, or when an airplane flies by outside the window (Andrew 1963*a*). In the wild, *L.macaco* coughs during excited locomotion, when mobbing the observer, when separated from the troop, when neighboring troops are approaching, and, in a variant, when settling to sleep (Petter 1962*a*). If the single sound has a single motivation, the motivation must be high arousal (or high stimulus contrast) in almost any type of situation (Andrew 1962*b*, 1963*a*). If we likewise assume a single function, it must be to raise other troop members to a state of high arousal. However, the cough occurs in well-defined contexts in the wild, so the troop responds with specific actions: speedy locomotion, alertness or scent-marking toward another troop, staring and tail swinging toward the observer, or, at dusk, echoing the "evening cry" of distant troops.

In the preceeding chapters, I have described particular displays under their function in the wild. This was partly for the readers' convenience: how to picture *L.catta*'s aggressive behavior, for instance, while referring elsewhere for a description of spats and stink-fights? The method chosen also seemed to me to be the most natural order: a study of social behavior in the wild should show how displays function in the natural setting.

It is now necessary to recapitulate with notes on the phylogeny and motivation of lemur displays.

OLFACTORY COMMUNICATION

Prosimians stand out among primates for the wealth and variety of their olfactory communication. Monkeys and apes have few specialized skin glands and almost no stylized olfactory behavior. Their only obvious olfactory signal is the scent of females in estrus (Marler 1965). Prosimians, however, use ritualized scent-marking. In this, they seem intermediate between other mammals and higher primates.

In this study, I observed four types of scent-making in *P.verreauxi*: throat, urine, fecal and genital with one associated behavior pattern, tail-lashing (Table II–7). *L.catta* had three sorts of marking: genital- and palmar-marking of branches and marking the tail with the wrist glands. With these three markings occurred two more behavior patterns: annointing the brachial gland with the axillary gland and tail-waving (Table III–5). *L.macaco* (Petter 1962*a*, Andrew 1964*a*) marks with its whole anogenital region and rubs its forehead on marked branches or in feces. *L.macaco* frequently mark other animals, although *L.catta* does not. All three species smell the urine and the anogenital region of troop members.

All the scent-marking described had a large visual component. Marler (1965) points out that the biochemical analysis of scent is a lengthy process; so behaviorists tend to describe olfactory signals in terms of the gestures of emission or deposition. These gestures, in the lemurs, are so highly ritualized that they must certainly function themselves as signals (Plates 15 and 16). In fact, the scents of *L.catta* and *P.verreauxi* are nearly imperceptible in the open to the human nose. I felt, watching a stink-fight or territorial battle, as though I were at a silent movie. I was unable to "hear" the main mode of communication but could follow it all through the actors' elaborate, exaggerated gestures.

Scent may well differ between subspecies of lemur and be one of the species recognition marks. It must also change with season and sexual condition. One *L.macaco* male in the breeding season became all too obvious to the human nose, although at other seasons he smelled little.

These diurnal lemurs apparently used scent-marking chiefly in social situations, not for orientation like the lorisoids. *Propithecus* scent-marked during territorial disputes, but their hesitation to invade each other's territories could be due to unfamiliarity or expectation of a battle with the owners as well as to old scent-marks on branches. It is, however, likely that nocturnal lemuroids, which have even more restricted ranges (Petter 1962*a*), orient by scent and use it to claim their own territories.

Scent-marking, in general, is difficult to assign a simple motivation or a single social function. I described it with territorial defense for *P.verreauxi* and agonistic behavior for *L.catta,* because we see its most striking examples in these two situations. When someone studies the mating season of *P.verreauxi*, he may find more intratroop agonistic functions of scent-marking.

Andrew (1964*a*) shows that scent-marking is evoked in greeting, threat, courtship, and by unfamiliar objects. The only all-embracing motivation then would be stimulus contrast. In the wild, I think we can narrow it down to self-advertisement in a wide range of situations. Self-advertisement, though, is hardly a traditional drive—it is only offered as a way of interpreting in human terms the common element of these situations.

TACTILE COMMUNICATION

The agonistic tactile signals of lemurs seem much like those of other mammals, including other primates. Poking with the nose, cuffing, and biting are common forms of agonistic contact. In lemurs they are relatively unritualized, except for the lifted hand in an intention cuff, which is a signal which may be understood even between genera of Lemuroidea.

The greeting gesture is a brief touching of noses. This type of greeting is more typical of other mammals than of higher primates, where the hand usually initiates contact (Marler 1965).

In contact and grooming, the hand again plays very little role, except to seize the other's fur (Bishop 1964, Buettner-Janusch and Andrew 1962). All lemurs groom each other in the same way, with upward scrapes of the tooth-scraper or tongue. In contrast to many other primates, grooming is generally reciprocal, with the two animals alternating or grooming each other simultaneously.

Grooming and contact, although different in form, have much the same function in lemur and monkey troops. They pacify and cement friendly relations within the troop. (See pages 159–63.) The behavior of the two lemur species observed are in accord with Marler's (1965) supposition that grooming is more frequent in aggressive species with a marked dominance hierarchy. *P.verreauxi,* with its peaceful family structure, grooms rarely, the irritable *L.catta* much more often. However, since grooming is reciprocal, we cannot usually see that the inferior member of a pair initiates grooming to pacify a superior.

Wrestling play also seems to be a high-intensity contact, or friendly relation. The gestures of play are remarkably similar among lemuroids. A pair of *Phaner* hung by their feet, sparred, and groomed at twilight, looking just like miniature *Propithecus.*

Mother-infant contacts, on the other hand, vary widely among Lemuroidea (Petter-Rousseaux 1964, and chap. VII). It would be interesting to see whether adult gestures of contact and grooming differ between those lemurs that build nests and those that carry their young.

VISUAL COMMUNICATION (TABLES II–5 AND III–7)

Visual gestures differ relatively little in lemurs, except for the gestures associated with scent-marking. Lemurs, as well as other mammals and primates up through the chimpanzee and gorilla, will stare in threat, swagger with head up in confidence, hunch over in a "resting attitude" (Andrew 1964) in submission, stand on their hind legs with raised hands in fight. There are only two stereotyped visual displays, beside scent-marking, that appear in just one family: the pendulum-swinging tail of *Lemur* and *Hapalemur,* and the head-jerk of *Propithecus,* both given toward ground predators.

The differences of color and pattern, however, change the gestalt of many displays between species, subspecies, or even males and females. The hunched, submissive posture may hide the bright face-mask of *L.macaco,* show the brown hood of *P.verreauxi,* or reveal a startling white neck patch in the red form of *L.variegatus.*

Bodily posture, of course, also varies between genera. However, changes in posture often seem to have no signaling function. *Propithecus* rolls up its tail or lets it hang straight without any apparent reaction from others. *Lemur*'s tail tends to hang down in the trees, counterbalancing its weight, and tends to be cocked up while walking on the ground.

Facial expressions have been extensively discussed by Andrew (1963*a*, 1964*a*). Three are particularly noticeable: the pursed mouth, the open mouth, and the grin (Plates 17 and 18).

Contraction of *m.* orbicularis oris purses the mouth into an "O." This increases resonance of the buccal cavity and so may accompany any very loud sound (Andrew 1963*a*). It appears in *P.verreauxi* with the roar toward hawks, and in *L.catta* in the "ow" of meow contact calls and in the howl. Andrew (1963*a*) concludes that the pursed mouth occurs more readily in *L.catta* than *L.macaco.* *Indri* represent far more extreme facilitation. They purse their mouths in alarm roars and almost certainly in their howl or song. They have evolved a special signal as well, for they evert their lips slightly, so the mouth shows bright red and round like flapper make-up of the 1920's.

The open mouth with teeth covered is a primitive mammalian threat expression which persists through higher primates (Andrew 1963*a*, Marler 1965). It appears in *P.verreauxi* and *L.catta* with play-bites, with exaggerated yawns which form part of more lengthy encounters, and with mobbing noises toward ground predators.

The grin in lemurs is especially interesting. Andrew (1963*a*) derives grinning in primitive mammals from "protective responses," including the "rejection response" to bad food in the mouth or strong odors. Wild lemurs grinned when smelling scent-marks, when threatened by superiors, and during all spat calls. The first syllable of *L.catta*'s meow is also accompanied by a grin. *L.catta* also grins more readily than *L.macaco.* Exaggerated chewing movements, which seem related to the grin, appeared during stink-fights and while nosing a (possibly odorous) chrysomelid beetle. This list certainly links the

reaction to odor with defensive social situations. However, it is difficult to be sure that lemur scent is really unpleasant to lemurs, and
even the beetle was being actively investigated. These observations
might still seem equivocal for deciding whether the protective or
emotional aspects of grinning are fundamental. There was, however,
one type of clear-cut rejection response. Juvenile *L.catta*, instead of
rejecting bad kily pods by smell, occasionally bit into a dry, eaten-out
pod. They then made exaggerated chewing motions, like adults in a
stink-fight. The connection between protective responses and defensive ones seems well established in lemurs. However, one human
baby (6 to 9 months), who consistently grinned, narrowed her eyes,
and turned aside her head in refusing food, has done as much to
convince me as all the lemurs of the truth of Andrews' thesis!

VOCALIZATIONS

SIMILARITIES WITH OTHER PRIMATES

The vocalizations of social Lemuroidea are not particularly "primitive." They have a large number of click calls, which may in fact be
an early form of primate communication, since they appear in lemuroids, lorisoids, and the infants of many higher primates. They
present, however, problems similar to those of higher primates—size
of repertoire and the same vocalization used in different situations.
Undoubtedly there has been much convergent evolution between
lemur vocalizations and those of higher primates, both of specialized
calls like the song of *Indri*, which resembles the song of the gibbon
(Andrew 1963a), and general ones like the graded grunts of *L.ma-
caco*, langurs, or baboons.

Many vocalizations have similar forms and functions in higher
primates, lemurs, and other mammals. These are very likely sounds
whose form is particularly appropriate for conveying information
about some type of situation (Marler 1965, Andrew 1964a).

One example is the alarm bark used by gorillas, chimpanzees,
baboons, rhesus monkeys, and langurs. *Indri, Propithecus*, and *L.ma-
caco* give shrill or roaring barks at the approach of a ground predator. *L.catta*'s mobbing bark or yap has a tonal quality to the ear, but
as the spectrograph in Andrew's monograph (1963a) shows, it has a
large component of noise. It is precisely the soprano barking or
yapping of a very small lapdog. As Marler (1965) says, barks usually
convey the message as quickly as possible without loss of energy.
(The roaring bark of *Indri* or *Propithecus* is, however, a major
display in which the animal sits immobile with muzzle upward.)

This sort of noise seems likely to evolve in social animals, where it is more important to alert the whole troop than to conceal the caller's whereabouts from predators. In lemurs as in dogs, barking is much more: it has become an earsplitting threat to the predator. Social facilitation of barking is so strong in lemurs that the troop even joins in synchrony after the first yap or roar of a bout. This kind of mob threat could only develop in social animals, and fairly well-protected ones at that.

Screaming in fright is also common among mammals, including primates. This probably derives from violent expiration combined with glottal constriction under the influence of the sympathetic nervous system. *L.catta* is exceptional in the facilitation of its shriek to flying hawks, but screaming has become facilitated in several higher primates, such as langurs and chimpanzees, in social situations.

Similarly grunting during locomotion and growling in mild aggression and infant clicks are common in lemurs as in other primates.

L. catta AND *L. macaco* COMPARED

Lemur species, however, naturally differ in their vocalizations. *L.catta* and *L.macaco* may be compared as a case study.

The first fundamental point is that it is possible, even fairly easy, to make a table like Table V–1. If we hear a call, it is often difficult to know the situation. However, if we know the situation, it is possible to predict the analogous vocalizations of either species. This again makes the point that, in the wild, the social function of a vocalization is usually limited to a number of predictable situations, even though an animal could potentially use such a call in a much wider range of circumstances.

L.macaco's grunt noises and cough noises are fine examples of vocalizations to low and high stimulus contrast with relatively little differentiation into distinguishable noises for different situations. *L.catta*'s clicks are also given in many situations, but at higher intensities they are quickly replaced by discrete "clear calls." The replacement of graded grunts in *L.macaco* by calls in *L.catta* is the most striking difference between them. It is not at all clear why this should occur. Calls are useful for troop contact in foliage which obstructs vision. *L.catta*'s meows are certainly used this way, but it does not seem obvious that their troops are so much larger than *L.macaco*'s as to make contact by locomotor grunts difficult.

Marler (1965) points out that, where several similar species inhabit the same region, they may need highly structured and relatively invariable sounds as species markers. He suggests that "the

TABLE V-1

VOCALIZATIONS OF *Lemur macaco* AND *Lemur catta*

L.macaco Vocalization [1]	French Equivalent [2]	*L.catta* Homologous Vocalization	*L.catta* Voc. in Same Situation	Occurrence
NOISES		NOISES	NOISES	
Click-grunt...........	on-on-on grognements	clicks, click-grunt	clicks, grunts	in fast normal locomotion, initiating locomotion.
Click-grunt...........	on-on-on grognements	clicks, click-grunt	–	friendly greeting
Long click-grunt.....	on-on-on grognements	clicks, click-grunt	–	friendly greeting at a distance
Short-grunt..........	on-on-on grognements	clicks, click-grunt	purr	mutual (genital) grooming
Short click-grunt.....	on-on-on grognements	clicks, click-grunt	single clicks	to strange objects
			CALLS	
Querulous grunt.......	on-on-on grognements	clicks, click-grunt	spat	when crowded or handled roughly by superior, after copulation
Bark.................	on-on-on grognements	clicks, click-grunt	light spat (yip)	driving off threatening inferiors, formidable strangers
Sharp call...........	on-on-on intense, aigu meum-meum	clicks, click-grunt	bark (yap)	to strange sudden stimulus, mobbing predators
—	"cri de canard"			*L.m.macaco* only, to other troops

Cough..............	crouii, cri aigu creee	explosive grunt ?	mew, meow meow meow	When isolated from troop in excited locomotion when frustrated by closed door when two groups meet
			bark (yap) shriek howl, meows	mobbing predator to hawk before settling to sleep searching for teat
Infant clicks........		CALLS		
CALLS				
Wail and moan.......		meow, mew	intense meow calls	long separation, subordinate in cage
Squeaks............	—	spat	deep spat	to strong attack of superior
Shriek.............	—	shriek	shriek	when seized by human
Infant squeaks......	—	infant squeaks	infant squeaks	separated from mother or roughly handled

¹ Andrew (1963a).   ² Petter (1962a).

close-knit, terrestrial or semi-terrestrial social grouping, and the relative isolation from numbers of cohabiting species . . . correlates significantly with the relatively ill-defined structure of most of the sound signals of the rhesus monkey, baboon, chimpanzee, and gorilla."

These two species of *Lemur* may well prove to be an exception, but they do not seem to confirm the hypothesis. *L.catta* is more terrestrial and has as highly structured social grouping as *L.macaco*. (It could even be argued that the facilitation of facial expression in *L.catta* corresponds to a *more* structured social group.) *L.catta* at the edges of its range overlaps with *L.macaco*, but if the calls were different in any way, they could be distinguished. An illuminating case here is *L.mongoz*, whose range is completely included within that of *L.macaco* and which on preliminary study seems to use grunts like *L.macaco*'s rather than *L.catta*'s wealth of clear calls (Petter 1962a, Andrew 1963a).

SIZE OF REPERTOIRE AND INTERMEDIATE VOCALIZATIONS

This brings up the whole question of size of repertoire and the distinctness of different sounds. Marler (1965) suggests that ten to fifteen vocalizations will be usual for most primates, with present lists ranging from seven to twenty-five. I listed five for adult *P.verreauxi* and fourteen for *L.catta*. Andrew gives eleven for *L.macaco*. Of these, all five *P.verreauxi* sounds are distinct. The *L.catta* sounds fall into five groups within which the sounds intergrade. *L.macaco* has four groups, but three are so rarely used that almost all its common vocalizations are variants of the grunt-cough continuum.

Marler (1965) and Bastian (1965) emphasize the importance of a variable grunt in higher primates as the precursor of human speech. Andrew (1963b) shows how facial expression in a species such as baboons may modify the grunt to convey information about the mood of displaying animal. As he points out (1964a), *L.m.fulvus* and *L.m.rufus* rapidly protrude their tongue in greeting, which gives a distinctive character to grunts of greeting. Whether modified by facial expression or by nonvisible changes in the buccal cavity, many higher primates produce graded types of sounds.

Bastian declares that it may have been necessary to proceed from discrete signals through variable grunts before arriving at the discrete but arbitrary signals of human language.

L.macaco reinforces this picture of the form of primate vocal communication but suggests that such a form might be evolved by any social primate, not by just the large, semiterrestrial, closer relatives of man. As in many other primate displays, this is a striking example of convergent evolution (Andrew 1963*a*).

Finally, it is possible to produce intermediate communications by combining sounds as well as by giving intermediate noises. An *L.macaco* chivvied by a superior makes querulous grunts, which at increasing intensities may end with squeaks. *L.catta* grunts at high intensities may end with a mew or a mew with a bark. The volume, and presumably the force of expiration, becomes louder, until with the kind of autonomic glottal closure described by Andrew (1963*a*) the vocalization changes quality.

Another informative combination of sounds is in troop choruses. The sifaka and roaring bark of *P.verreauxi* and the meows, yapping bark, and scream of *L.catta* are all given when another member of the troop starts the same call. Further, the mood of the entire troop is easily understood by the combination of calls. The relative proportion of clicks, grunts, mews, and meows shows how fast the troop is moving, whether it is starting movement, in continuous progress, or changing direction. The timing and proportion of clicks and yaps, or growls and sifakas, reveal the extent the troop is disturbed by a predator. This, in turn, reinforces the existing mood in each animal, like the contagious cold silence, titters, or guffaws of a theater audience.

THE NEED FOR PHYLOGENETIC STUDY

This chapter has noted a number of ideas about the nature of primate sound. Andrew's concept of stimulus contrast and Marler's emphasis on the importance of gradations and intermediates between displays have been set against the common observation that the function of a display in normal social contexts is usually quite definite and precise. Comparison of *L.catta* and *L.macaco* showed how two, equally arbitrary, sets of signs could be used in similar circumstances, one set consisting of intergrading grunts, the other of different clear calls.

Similar phylogenetic comparison among all the Lemuroidea is missing. Only with much more detailed study could we begin to speculate which vocalizations have diverged as species-isolating mechanisms and which ones may be adapted in form to serve particular social functions. All the other evolutionary questions raised by a series of homologous and analogous structures remain unanswered.

On the simplest level, many subspecies differ in their calls. The sifak of *P.v.coquereli*, for instance, differs from that of *P.v.verreauxi*. Petter (1962*a*) gives sound spectrographs of the different coughs of

L.m.macaco and *L.m.fulvus.* On the species level, we find differences in quality; *P.d.diadema*'s sifak was transcribed by two independent groups of observers as "vouitch" and "vooch" (Petter 1962*a*, Maxim and Boggess, personal communication). We also find facilitation of different types of vocalization, as in *L.macaco* and *L.catta.* On the generic and higher levels, there will be many more cases of homologous noises with various functions, as, for instance, I unhesitatingly relate the roaring bark of *Propithecus* to the "klaxon" alarm call of *Indri,* or the howl of *L.catta* to the song of *Indri.*

Lemuroidea are very vocal primates, in fact, extremely noisy ones. The klaxon call of *Indri,* the howl and bark of *L.catta,* the roaring bark of *Propithecus* all resound through the woods. Petter (1965) describes the "intense roars and clucks" of *L.variegatus,* which are answered in kind by distant troops (Petter 1962*a*). *Lepilemur* also makes intense calls at night (Petter 1962*a*). *Phaner* squawk in chorus and are echoed at once by neighboring groups. When *Phaner* troop members rejoin each other, they make clicks like amplified static. Even the solitary aye-aye produces a "grinding noise like two metal sheets . . . rubbed together, which is answered periodically by a similar cry from another animal in the distance." Although Petter (1962*a*) and Andrew (1963*a*) have recorded many lemuroid calls, there is obviously still a marvelous study here for anyone who will wander clockwise around Madagascar with a tape recorder.

Chapter VI

The Importance of Territory

All the social Lemuroidea known have sharply localized, stable home ranges and territories. This includes *Propithecus verreauxi coquereli* at Ankarafantsika, *P.v.verreauxi* at Berenty and Lambomakandro, *L.catta* at Berenty, and *L.macaco* at Nosy-bé and Lambomakandro (Petter 1962*a*; Boggess and Smith, personal communication). Probably *Indri* and *Hapalemur* troops at Perinet and *Phaner* at Lambomakandro are equally localized geographically (Maxim and Boggess, Boggess and Smith, personal communications). The semisocial *Lepilemur* lives in small areas or neighborhoods made up of several animals, each with its own minute home range (Petter 1962*a*).

In the forest at Berenty the home ranges of *Propithecus* and *Lemur* troops overlapped with those of neighboring troops, but each had a central territory which was successfully defended against other animals. Territorial battles took place in the zone of overlap, so that every part of the forest was defended by one troop of each species or lay on a narrow disputed boundary line. This differs from the situation diagrammed, for instance, in Bourlière (1964) in that there is no neutral ground.

Many primates, like redtails (Haddow 1952), langurs (Jay 1963) and gorillas (Shaller 1963), seem to overlap and mingle without any attempt to defend territories. Others, such as gibbons and howler monkeys (Carpenter 1934, 1940) sing to each other from a distance, while Indian rhesus even fight physically (Southwick 1962). Social Lemuroidea, then, are among the primates which have marked territorial behavior, although it may not involve physical contact. The first sign of dissolution of the *Propithecus* Red Head troop was its inability to defend a territory.

One surprising thing about lemuroid home ranges is their minute size. The home ranges of *P.verreauxi* troops at Berenty were only 2–2.5 hectares (5–7 acres, 0.01 sq. mi.), while a troop of *L.catta* occupied only 5–7 ha. (0.025 sq. mi.). *P.verreauxi* at Ankarafantsika had ranges of about 0.5–2.0 ha. (Petter 1962*a*). *L.macaco* on Nosy Komba had larger ranges, one troop as much as 40 ha., but this included degraded second growth and plantation. A troop which lived in a remnant of forest was confined to 6 ha. (Petter 1962*a*).

Among other primates, the forest-living colobus, howlers, and gibbons have troop ranges reported to be 0.06 sq. mi., 0.5 sq. mi., and 1.0 sq. mi., respectively. Savannah-living baboons may range over 15 sq. mi. The highest population densities known are one hundred thirty-six animals per sq. mi. for Barro Colorado howlers (DeVore 1963*a*) and two hundred and twenty-one sq. mi. for forest-living Hanuman Langurs (Sugiyama 1964). By contrast, the Berenty populations of *Propithecus* and *Lemur* were respectively equivalent to six hundred fifty and eight hundred thirty animals per square mile, although this assumes uniformly favorable forest. (Squirrels, which somewhat resemble *Lemur* in food and locomotor habits, often exceed this density [Mohr 1947].)

Such geographical localization has a number of biological implications. First, it increases the degree of inbreeding among social lemurs. Next, it provides a neighborhood in which all the animals are acquainted with each other. Finally, it is an easily studied unit which defines the ecology of each troop, including their relations with other lemur species.

INBREEDING, POLYMORPHISM, AND SUBSPECIATION

The degree of inbreeding in social Lemuroidea must be extremely high compared with that of farther ranging mammals. This appears often in family resemblances. For instance, the Blaze Nose troop of *Propithecus* had a mother, a juvenile of 1963, and an infant of 1963 with a sharp blaze down the nose. The solitary male who lived just southwest of the Blaze Nose troop and one member of the troop just northwest had the same mark. I did not see blazed noses on any other animals of the Mandrary forests. Other recurring marks were the widow's peak in three males of Widow Peak troop and a capline jagged up to the right in six of the animals of the Bevala melanistic troop.

The two troops at Bevala deserve special mention, or even a special study, as extreme examples of lemur inbreeding. These *Propithecus* have been isolated for about 20 years after the surrounding woods were cut to plant sisal. There were in 1964 a total of eighteen animals, one troop of eight and one of ten. (Ten is the highest troop size recorded for *Propithecus*, well beyond the normal range.) In the troop of eight animals, there were seven ordinary white ones and one heavily melanistic one as described (Hill 1953) for *Propithecus verreauxi majori* (Plate 19). In the troop of ten, there were four white and six melanistic. Melanistic animals varied in depth of coloration. Some had light gray backs, some medium brown. Their forearms and thighs were chocolate on the leading edge, but in some this chocolate color extended down to the knee and in some only over the proximal part of the thigh. In a few animals, the cap continued down in long sideburns to dark chocolate cheeks, encircling a strip of white running from the brow around the edge of the black face. In other animals, the cheeks had only faint grayish patches on them. One melanistic female had a normal white infant.

As we have seen, *Propithecus* may be able to breed in their first year, though more probably not until their second year. In a troop isolated for 20 years, it seems that the conditions necessary for genetic drift are fulfilled. Of course, it is impossible to prove that genetic drift has occurred. For whatever cause, there has been a steady increase in the number of melanistic animals at Bevala, 20 years ago the troops were either entirely or almost entirely white (A. de Guiteaud, personal communication).

The troops at Bevala illustrate clearly two points about variation in *Propithecus* and other Lemuroidea. First, there is a good deal of individual variation; each animal's face and markings are different. There is such sharp polymorphism between white and melanistic animals that these were previously called two different subspecies. On the other hand, the troops also illustrate that there is a great deal of inbreeding which leads to marked family resemblances, for example, the line of the cap being the same or almost the same in most of the members of the larger troop. The capline is a very widespread variation, so much so that *Propithecus* can usually be identified at once by the shape of their brow, although members of a family may be confused. The melanistic form appears in other forests, notably at Ifotaka and on the western side of the island at Lambomakandro.

Inbreeding, by reducing the rate of gene flow between neighboring populations, should produce a large number of small, semi-isolated populations which differ from each other very markedly, but within which the animals resemble each other very closely. In *Propithecus*, however, there may have been a selective pressure for individual variation within a population, since the animals must recognize individuals and behave differently toward each one. If it is argued that *Propithecus* troops are so small that selection for individual recognition is unlikely (Andrew, personal communication) it seems at least possible that individual recognition and troop life have relaxed some of the pressures for uniformity in a species. These two opposing tendencies seem to have produced in *Propithecus* a situation such that, in one race, the individuals may vary a good deal; but where there is a geographical boundary dividing populations, the subspecies again differ sharply from each other, so that morphologically different populations often appear on either side of a major river.

There are about nine subspecies of *Propithecus*, divided between two closely related species. The subspecies are separated on geographical grounds and on fur color, not on skeletal characteristics. This becomes complicated because, as we have seen, there may be sharp polymorphism in fur color even within one troop, while intermediates exist between several of the subspecies.[1] Although a large number of specimens of *Propithecus* have been collected, the geographical origin of most of them is uncertain. We have no clear picture of the distribution of color types in *Propithecus* and certainly no information about troops which may contain more than one color of animal. Therefore, it is difficult or impossible on present information to sort out the races of *Propithecus* more than has been done by Hill (1953). However, the behavioral data suggest how this animal, in which the individuals are highly localized and in which inbreeding must be very common, could develop a high degree both of local individual polymorphism and of geographical color variation.

The same remarks apply to *L.macaco,* which has at least eight subspecies scattered around the island. *L.macaco* forms differ from each other more sharply than *Propithecus* do, but no poylmorphism has been reported within a troop. One special point which arises in *L.macaco* is that we know many of the forms have different chromosome numbers (Table I–3). In fact, one subspecies, *L.m.fulvus,* itself seems to have two different chromosome numbers, $2n$ being either 48 or 60. This becomes less puzzling with the field observation that there are two widely separated populations of *L.m.fulvus,* one in the eastern rain forest (map II) in the region of Perinet and one in the west, including the forest of Ankarafantsika. The eastern animals are slightly darker in body and head coloring with slightly lighter white beards. The western animals have rather tan or cream-colored beards and a grizzled lighter gray body and head. It may eventually be decided that there are cryptic species which should be separated on a specific level. However, zoo records exist of hybridization between

---

[1] At Berenty there were a few normal *P.v.verreauxi* which had a dark gray patch on their back and a heavy orange cast on their shoulders and upper thighs. This approaches the coloration of *P.d.diadema,* which have a charcoal gray back and orange arms and legs.

L.m.macaco and other subspecies (Moog 1959); so it seems more likely that we have here an example of extreme subspecific radiation,[2] or even chromosome polymorphism between individuals, such as in shrews (Ford, Hamerton and Sharman 1957; Meylan 1960).

L.catta seems to be in a somewhat different situation, since it is clearly a species with no subspecific forms and is confined to only one geographical area of the island. Even *L.catta*, however, has family resemblances between members of a troop. It is particularly difficult to say why one species varies more than another or which ones would benefit from isolating mechanisms because of the large number of recently extinct forms which must have influenced the evolution of the present species. Interpretation of the fossils is also difficult: *P.verreauxoides*, for instance, might be the same as the giant *P.v.verreauxi* seen by J. DuPray in the upper Mandrary headwaters (J. DuPray, personal communication). All we can say here is that inbreeding would favor the radiation and differentiation found in lemurs.

THE NEIGHBORHOOD

The second major implication of geographical localization is that animals of neighboring troops are acquainted with each other. Animals are, of course, more familiar with members of their own troop. However, *L.macaco* (Petter 1962*a*) and *Propithecus* engaged in intraspecific territorial battles in which the animals leaped toward each other and occasionally even came in contact, and in which the animals of each troop scent-marked the disputed territory and smelled each other's scents. It seems very likely that they learned both the scent and appearance of members of adjacent troops. This probably applies more to *Propithecus* and *L.macaco* than to *L.catta*, in which the groups rarely come in contact. Petter (1965) describes an *L.macaco* dispute in which animals actually cuffed each other and pulled each other's hair. However, my observation that males of an adjacent troop attempted to mate with a *L.catta* female makes clear that there may be considerable contact at times between neighboring animals.

When the *P.verreauxi* Red Head troop broke up, two members of the troop began to move with the solitary male, Sideburn, who lived in the adjacent territory. Thus, it seems possible, if not likely, that

---

[2] Analysis of serum proteins and hemoglobins distinguishes *L.catta* from *L.macaco*, but not between subspecies of *L.macaco* (Goodman 1962, Buettner-Janusch and Buettner-Janusch 1964).

new *P.verreauxi* troops may be formed by adult members of neighboring troops who are already acquainted with each other.

The behavioral result is that the world of *Propithecus* is not confined to the other members of the same family but is extended through repeated acquaintance to nearby animals. Since all the animals maintain a stable geographic relationship, it is possible for the members of the neighborhood to know the individuals which surround them (their sex, their relative tendencies to aggression) and to use this knowledge in the eventual formation of new troops. The knowledge is also used in ordinary interactions between troops. In a very crowded area, it is possible for animals to feed in sight of members of another troop without aggressive interaction. They may even cross a corner of another troop's defended territory indicating by their speed and direction they are not provoking territorial dispute. It would be very interesting to introduce one or more strange *Propithecus* into such a settled area and see whether they would be treated with much more hostility than the residents. DeVore and Washburn (1963) state that neighbor baboon troops do recognize and tolerate each other and are nervous around strange troops. One interesting variation appears in *L.m.macaco* (Petter 1962*a*), whose troops may occupy and defend territories during the day but join to sleep in the same dormitory trees at night.

In short, *Propithecus* or *Lemur* social behavior does not stop with the interactions within the families; there is a reserve of less likely, but still possible, interactions between the troop members of a geographical neighborhood.

THE ECOLOGICAL UNIT AND POPULATION GROWTH

Animals that maintain a year-round limited territory are very convenient study objects for the ecologist. An ecologist may try to calculate the amount and variety of food necessary for an animal, the amount of water or even the number and kind of sunning sites. Animals like the social Lemuroidea, which maintain a circumscribed territory, make these calculations for the ecologist, for each troop attempts to defend an area sufficient to its needs.

It is another question what these needs are. Presumably the territorial size limit is set behaviorally by something like the physical activity of the animal or by its capacity to remember geographical landmarks. Only in exceptional circumstances would actual hunger or other fundamental drives force a troop to enlarge its range. However, natural selection would lead a species to range over and defend the optimal areas for its survival. Where population pressure is great, as in a natural reserve surrounded by sisal fields, we would expect range and territory to be the minimum size compatible with species survival in a stable population. There is no way of knowing if the Berenty populations are stable, but *P.verreauxi* territorial sizes there seem comparable to those at Lambomakandro and Ankarafantsika (Boggess and Smith, personal communication; Petter 1962*a*). The theoretical ecologist likes to think of the environment as divided into a mosaic of supplies utilized by each animal and each species of animal (Hutchinson and MacArthur 1959). Animals which have a year-round living territory, not merely breeding territories, divide the environment sharply into a geographical mosaic for themselves.

During 1963 and 1964 at Berenty, food did not apparently limit *Lemur* or *Propithecus*. They fed on fruits and foliage and made no appreciable difference in the amount of either. Their habit of moving to different parts of the range every 2 or 3 days, of course, spread the amount of damage and did not localize it in one tree or one area. However, I did not have the impression that there was scarcity of food. Neither did there seem to be a scarcity of drink. *L.catta* does descend to the river to drink, but many troops have ranges which are not adjacent to the river, and I did not see these troops cross over into strange ground in order to reach the water. There also seemed to be a great many sunning and sleeping trees. In short, the territories and home ranges occupied by *P.verreauxi* and *L.catta* that I studied seemed to amply satisfy all the animals' obvious requirements.

In the relatively wet years of 1963 and 1964, the nine troops of *P.verreauxi* which inhabited the 10 hectares I studied increased by about 10 per cent in population.[3] In the same area and time, *L.catta* troops 1 and 2 increased by about 25 per cent. It does not seem, therefore, that there are immediate and current population controls in force. An alternative possibility suggested by T. Rowell (personal communication) is that primates ordinarily build up their populations at an extremely high rate. Then they either split up into new troops which colonize unused land or suffer a catastrophic population crash which removes a large proportion of the animals—probably the crash is caused by epidemic, not by predation. There are few primate data which would confirm or refute this hypothesis. The censuses of howler monkeys on Barro Colorado Island are suggestive, for the population rose from 398 to 489 during 1932 and 1933, dropped (possibly from yellow fever) to 239 in 1951, and rose again to 814 in 1959 (Collias and Southwick 1952, Carpenter 1962).

---

[3] This allows for the breakup of the Red Head troop and its replacement by the Peaked troop.

The question of population growth will have to remain open, but it would be well to attempt to resolve it by studying territorial primates in which the ecological space of a particular troop corresponds with an easily measured geographical space.

INTERGENERIC RELATIONS

Finally there is the question of relations between the various forms of Lemuroidea, and whether there is any benefit or disadvantage to one species from its association and sharing of territory with others.

Propithecus and *Lemur* are certainly aware of each other; even when feeding in the same tree in apparent indifference, they kept at least a meter's distance between individuals. *Propithecus* would often glance nervously at nearby *L.catta*. The incidents that I described as examples of one kind of lemur teasing the other would probably occur more often in a very crowded situation and provide a measure of the frequency of the other animal's presence (Wynne-Edwards, 1962). Although *Lemur* are generally less intelligent and curious than higher primates (Jolly 1964a,b, they did at least show a certain curiosity about other primates of the forest; for instance, one *L.catta* approached *Lepilemur* and two juveniles came up and bit C. H. F. Rowell's shoelaces.

One interaction of clear mutual advantage was in predator protection. Each species responded to the predator calls of the other and gave appropriate responses. When *Lemur* gave the hawk scream or *Propithecus* the hawk roar, troops of the other species would take cover and look up into the sky. When either began their mobbing calls, the sifaka or the bark, the other species would leap up away from the ground and look downward for the predator. They did not, however, join in mobbing unless they also saw the predator. This was clear when *L.catta* troop 1, which accepted me, met strange *Propithecus*, who began to sifaka. At the first alarm the *Lemur* would leap up, look around, see me, and almost contemptuously return to the ground.

The ecology of *Propithecus* and of *Lemur* is very similar. They do divide the habitat to some extent. *L.catta* at Berenty spent about 15 per cent of its time on the ground, *P.verreauxi* only 5 per cent but *P.verreauxi* spent slightly more time in the highest branches. In their food, though, they were nearly identical. Seventy per cent of the plant species which they ate were eaten by both animals. However, 80 per cent (*L.catta*) or 95 per cent (*P.verreauxi*) of the observations of feeding were of those species eaten in common. Their food was divided between leaves, fruit, and flowers in almost exactly the same way. In about 25 per cent of the observations, each lemur ate leaves; in about 70 per cent, fruit, and 5 per cent, flowers. I do not agree with Petter (1962a) that the direction or size of branches has much influence on the ability of *Propithecus* or *Lemur* to feed from the branches.

Furthermore, in the forest of Lambomakandro, which, to the eye, offered no greater diversity of plant food than Berenty, these same forms share their ranges with a third lemur of similar habits, *L.m.rufus*. In the extreme case of the eastern forest, *Indri, Propithecus*, three species of *Lemur*, and *Hapalemur* coexist with no obvious niche differentiation in either food or branch height (Petter 1962a, b), although there may be preference for hilltops or ravines (Maxim and Boggess, personal communication).

Most dicussions of theoretical ecology concern the limiting conditions of competition between two species. That is, they cover such a large time span that two similar species must find a way to differentiate their requirements for shared resources or else compete to one species' extinction (Hutchinson 1957, 1959, 1965). The differences in requirements may be qualitative, for instance, a tendency to feed on particular foods, in which case the species are relative "specialists." If the species share preferences for the same sorts of food, and if they are jack-of-all-trade species which overlap broadly with each other, they must separate in space, although perhaps only by feeding at different heights of the same tree (MacArthur and Levins 1964).

Sympatric Lemuroidea have some obvious differences—in size, in water requirements, in tendency to feed on the ground. However, one aspect of their ecology is difficult to fit into the theories; that is, their behavior and ecological preferences can change within very wide limits.

All genera studied are extreme jacks-of-all-trades. *P.verreauxi* and *L.catta*, for instance, eat an entirely different range of food plants at Lambomakandro and at Berenty, and the *L.m.collaris* at Berenty shared *L.catta's* diet. Under changing climatic conditions, both genera could change their food plants. If there were increased competition from one genus for a particular food, the other genus could easily shift to a wide range of other foods, almost as long as there was any remaining vegetation.

This does not, of course, negate the basic theory. What it does mean is that two or more jack-of-all-trade species, which are already somewhat different, might coexist, sharing practically the same niche in times of abudance and differentiating their niches at need and possibly in several ways when resources are scarce. All the observations were made in times of abundance. In bad times, the differences between species and genera would diverge far more sharply as each was forced to exploit its own specializations to the utmost. Niche differentiation thus would only be crucial, and need only appear, at intervals during the unpredictable dry years in the south, as when, during 1943 and 1944, drought and famine killed many of the human population. One minor example of this differentiation took place in 1964. A typhoon shook down many kily pods. *L.catta* ate these from the ground more readily than *P.verreauxi;* so they formed a relatively larger part of *L.catta's* diet than the year before.

It is also quite possible that teasing could be used to drive one genus from a place where the other was feeding, so that competition might even be shown in behavioral interactions.

CONCLUSION

Lemuroid territoriality sharpens and makes precise many questions of their evolution and ecology. The degree of inbreeding is a crucial factor in any species' evolution (Mayr 1963). The extreme geographical stability of the Lemuroidea means that they have an extreme degree of inbreeding, which leads to the supposition that polymorphism and subspeciation in Lemuroidea are related to this inbreeding. In any social animal, it is important to know how many individuals one animal can recognize and must deal with. The territorial system means that a lemur has outside its troop a neighborhood of individuals with whom it is acquainted and with whom it may occasionally interact. The territory is an ecological unit defined by the animal; it supplies all the animal's needs. Not only do troops of one species limit or extend themselves to a given area and food supply, but the sympatric species or genera also are so limited. Thus, the community of all the Lemuroidea takes on the neat spatial structure dreamed of by ecologists. However, each lemur eats such a wide range of plant food that their niches may largely overlap in seasons of abundance and be differentiated only in times of scarcity. Therefore, to understand their ecology, it may be necessary to record a long-term temporal pattern as well as the spatial one.

Chapter VII

Lemuroidea and the Evolution of Primate Social Behavior

The first, and fundamental, similarity between lemuroid societies and those of other primates is that they live in groups with several adult males, and that an infant of either sex is likely to remain for life in the group of its birth. There are exceptions to this rule, of course, in several species of primates (marmosets, gibbons, and chimpanzees) (Hill 1957, Carpenter 1940, Goodall 1963), and there are a few other mammals which live in such societies. Still, this type of society is rare in other mammals (Andrew 1964a), while it is the common pattern in all three great branches of primates.

New and Old World monkeys may have evolved separately from a prosimian ancestor, developing complex social life and intelligence in parallel with each other. The social Lemuroidea certainly have had no ancestor in common with New or Old World monkeys since the prosimians of the Eocene or even the Paleocene (Simons 1963, 1964). Yet all three branches form highly cohesive social groups with the same basic kind of group structure in most of the species studied within each branch. Clearly the ancestor of all these animals was either social itself or else (more likely) had certain traits of behavior which, when its descendants were exposed to the same selection pressures, led them to form the typical primate kind of society.

The dispersionary mechanism in many animals is aggression. especially between the males. One mechanism of achieving a permanent society, then, would be to reduce aggression to the point where there were few conflicts or to the point where a defeated animal would be willing to remain as a subordinate in the victor's group and a dominant animal tolerated the company of such subordinates. A second, complementary mechanism is to increase some cohesive force between the animals such that they have a need to remain in the company of others. Binding forces which have been suggested in primates are sex, troop-infant bonds, and contact-grooming behavior. Both the reduction of aggression and the increase of cohesive forces, of course, usually operate at once: if there were no aggression but the animals were indifferent to each other, they would simply wander apart, but a strong impulse to stay together combined with full-scale aggression would lead to serious or fatal conflicts.

AGGRESSION

The level of aggression and the means of controlling it vary enormously among primates. Some genera of each branch seem to have suppressed the expression of aggression within a troop almost entirely—the howlers (Carpenter 1934), the gorillas (Schaller 1963). In others—particularly baboons (Washburn and DeVore 1961) and macaques (Altmann 1962a)—aggressive interactions may be common. Aggression among the same species in different environments also varies immensely: baboons (Rowell, in press) and vervets (Gartlan, personal communication; Struhsacker, in press) in lush Ugandan forest may have far fewer aggressive interactions than those of the Tanganyika plains. Macaque species may bite and wound while fighting (Southwick et al. 1965; Simonds 1965), but Washburn and DeVore (1961) stress that baboons have checks and restraints: their frequent expressions of power and antagonism were only once seen to lead to injury (Hall and DeVore 1965). As Andrew (1963b) has said, an animal with as much latent aggression as the baboon is under great selective pressure to maintain elaborate and exact means of communication to avoid real fights.

L.catta and probably P.verreauxi have a rather different system from higher primates. They have a very high peak of aggression during the breeding season, when low-ranking males challenge their superiors, and comparative calm for the rest of the year. Although many other primates have an annual breeding season (Lancaster and Lee 1965) and an increase in aggression around estrus females (DeVore and Hall 1965), none seems to have such dramatic changes as L.catta. It might be possible to consider the lemur pattern as a lower stage of socialization in which males fight over estrus females with almost the abandon of solitary animals and group structure could only be preserved by the shortness of the breeding season. However, this fighting does not seem to drive males from the group. Most of L.catta and P.verreauxi males remained with the same troops from year to year, the obvious losses being probably due to death from old age, out of the breeding season. The only bites visible out of the breeding season in either genus were old notches in the ears. The roughly 5-cm slashes received by *L.catta* males clotted almost immediately, and in a few weeks grew new fur; they seemed scarcely to impede the animals' progress. Thus, in spite of the fighting, there were few lasting injuries. It is therefore safest to say Lemuroidea merely have a different social structure which concentrates overt aggression at one period.

Within the lemur pattern, as in other primates, the total level of aggression differs between genera. The placid *P.verreauxi* practically never dispute, while *L.catta* even at the calmest seasons will have several minor spats an hour. In both species, however, these mild disputes over precedence on a branch or the right to groom an infant are at such low intensity they could hardly threaten the structure or cohesion of the troop. These two genera recall differences between howlers and *Cebus* in the Panama forest, *Colobus* and baboons in Uganda, even langurs and macaques in India. It is apparently common for species which share the same trees to have widely different levels of aggression.

One peculiar aspect of aggression in Lemuroidea is the relative dominance position of males and females. *P.verreauxi* and *L.macaco* (Petter 1962*a*) are the only known primates where the male-female ratio in a troop seems commonly greater than one. The troops I saw of wild *L.catta* and captive *L.macaco* are the only primates where all females can be said to be dominant over all males (Jolly, in press). Male *Lemur* are far more aggressive among themselves than females, but they are inhibited from attacking the females and afraid of the females. Females, on the other hand, show no fear of the males and will cuff or take food from the dominant male of the troop—or even chase him in unprovoked attack. In general, males are more status conscious, guarding their rank with glance and posture as well as overt attack. A dominant *Lemur* male is an imperious and bullying animal; a dominant female is simply bad-tempered enough to pick fights. The fact remains that troop 1 and 2 females won all their disputes with the males.

It remains to be seen whether this peculiar dominance structure in *L.catta* and captive *L.macaco* is related to the peculiar sex ratio observed in *P.verreauxi* and *L.macaco,* and whether it appears in still other Lemuroidea.

SEX

Zuckerman (1932, 1933) gave great impetus to the study of primate behavior when he suggested that year-round breeding was the original cohesive force of primate societies. Recently his views have been much disputed (Washburn and Hamburg 1965*a*). In savannah baboon troops, fighting increases and an unstable dominance hierarchy may be threatened when a female comes into estrus. Consort pairs withdraw to the periphery of the troop to avoid conflict with others. Lancaster and Lee (1965) attack Zuckerman's basic data and summarize recent studies that show baboons, macaques, and langurs in widely different areas have an annual season of matings and births. Sade (1964) shows that rhesus males are not even physiologically capable of breeding at all seasons.

Social Lemuroidea offer one final objection to the original theory. They breed seasonally and presumably always have done so, never developing a year-round monthly estrus cycle. Yet they share the basic primate form of year-round life-long social organization. Such organization must then have evolved among them with other cohesive social forces than permanent sexual attraction.

Lemur sexual behavior is unique because the season is so very short. I believe that *L.catta* troop 1 mated for less than 2 weeks in 1964, judging by the genitalia of the nine females. Observed mating and fighting lasted only 1 week. Each female seems to have been receptive for no longer than a day. Neighboring troops were sexually active at the same period, but not apparently longer. In 1963 both *L.catta* and *P.verreauxi* births were concentrated in a brief season, over perhaps no more than 2 weeks each. Cowgill *et al.* (1962) also reported some synchronizing of estrus in captive *L.macaco*.

The *Lemur* local breeding season must be among the shortest in mammals. Of breeding seasons listed in Parkes (1956) only the eastern skunk, *Mephitis mephitis*, may have a season as short as 2 weeks (Wight 1931). In skunks, as in Lemuroidea, there must be pronounced social or environmental synchronizing mechanisms, because nine of twelve captive females littered the same day.

Estrus in *Lemur* females could be synchronized by several means. It seems to me very likely that there is a great deal of social and olfactory stimulation. Olfactory stimuli alone are known to induce estrus in mice (Parkes and Bruce 1961) and to play a role in sexual attraction in most mammals. There is ample opportunity for lemur to become aware of each other's scents in mutual grooming and in genital-marking of branches, which increased sharply in the breeding season. Other social stimulation increases as well. In troop 1, the observed number of friendly contacts climbed from about two per hour to a peak of almost five and a half per hour 2 weeks before mating. Total spats in the troop, which were heard even when not observed, climbed from about six per hour to a plateau of about twenty per hour for the first 3 weeks of the month before mating;

they fell off as they were replaced by more serious stink-and jump-fights. I had the impression of spiraling excitement in the troop, which was communicated from animal to animal, each responding to the other's behavior as well as to its own physiological needs. In rhesus troops, differences in sexual and social relations may lead neighboring troops to different peaks of mating (Koford 1965).

Another possible intrinsic mechanism is age (T. Rowell, personal communication). All the young are born at approximately the same time, and all breed at the same age. Daughters might thus have the same season of estrus as their mothers. This has been seen occasionally in hamsters and in Norway rats (Calhoun 1963), but it is hardly stable or accurate enough for the precise timing observed.

There must also be an external trigger, for the 1963 birth peaks of *P.verreauxi* and *L.catta* at Bevala, 25 m from Berenty over bare sisal fields, corresponded approximately to those at Berenty. The obvious trigger is day length. Madagascar lies between 12° and 26° S. Lat.; so the days vary appreciably during the year. At Berenty, time of sunset and darkness advanced by almost an hour during March and April.

Cowgill *et al.* (1962) made a different suggestion. Of fifteen matings of two *L.macaco* females in their laboratory, thirteen fell within 5 days before or after a full moon, the remaining two within 9 days of a full moon. The estral reddening of these females lasted from 2 to 6 days.[1] Their first three estral periods, after they were moved to the northern hemisphere, did not coincide with the full moon; after this, six of seven estral periods overlapped a day of full moon. Cowgill *et al.* (1962) suggest that some effect of the moon may be widespread. In fact, almost any mammal must register the cycles of a tropical moon. Harrison (1952, 1954) has already pointed out that Selangor rats breed more often just before the full moon.

Evidence from the wild neither proves nor disproves the moon hypothesis. *L.catta* troop 1, on the evidence of genitalia, probably mated in the week of the full moon, April 9, 1963, and during the 2 weeks preceding the full moon of April 26, 1964. Troop 2 probably mated slightly later in 1964 with females in full estrus during the week before and at least 4 days after the full moon. Observed mating in troop 1 took place from 5 days before to 2 days after the full moon. This fits the hypothesis adequately; but if the mating season lasted about two weeks, there was a 50 per cent chance in any year that full

---

[1] Cowgill *et al.* (1962) report an *L.m.albifrons* female, having black genitalia, does not show visible oestrus. It now seems likely to me that any *Lemur's* genitalia normally swell at estrus, revealing an unpigmented center which flushes, like that of *L.catta*.

moon would fall during the breeding season. One confirming point is that five of the nine females grew pink about 1 month before breeding, around the preceding full moon of March 28 (Fig. 7).

In short, the lunar hypothesis remains possible. Positive, unequivocal laboratory results could prove it, and two seasons' field work on *Lemur* (or, I believe, *Propithecus*) could either disprove it or make it highly plausible. (Negative laboratory results would be irrelevant unless natural conditions were almost perfectly reproduced.) It seems important enough to be worth the effort. They might even indicate that for *Lemur,* as for the Arabs, Madagascar is the Island of the Moon.

The next question is, Through what steps did this synchrony evolve? It seems reasonable to suppose that the Lemuroidea were originally polyestrus within one season. Among prosimians, cycles of 21 to 60 days are found in *Tarsius* (Hill 1955), the Cheirogaleinae (Petter-Rousseaux 1962), and many genera of lorisiformes (Manley, personal communication). This condition then is common to all three branches of prosimians, at least in captivity. *L.macaco* subspecies in captivity also were polyestrus for a few months (Cowgill *et al.* 1962), and in the wild five females of *L.catta* troop 1 had a minor reddening of the genitalia about 1 month before the troop mated. It seems very likely that seasonal polyestrus was the primitive condition in primates, and that *Lemur's* synchronous estrus is superimposed on the basic condition.

The final question is, What is the function of this synchrony? There are several possibilities, none of which seems an adequate answer. *Lemur* males and females both seem to be sexually seasonal. This is also true of the Cheirogaleinae (Petter-Rousseaux 1962). In Lemuroidea there is little obvious difference in the size of the male genitalia from season to season, but the males were only seen to have erections beginning with the week before mating until the end of the mating period. Thus the males and females of a troop could be synchronized to be sexually potent at the same time. Another possibility is that male aggression is so dangerous that if it were continuous over several months, males would kill each other or break up the troop by driving members away. A third advantage might lie in the very sharp birth peak. If a predator, such as a hawk, could take only young, newly independent lemurs, the hawks could only seize a much smaller proportion of the young than if all young passed through this brief vulnerable stage at different times (Ashmole 1963). Finally, in an *L.catta* troop, infants are groomed as much by the mothers of other infants as by their own mother; so it might be an advantage to have many mothers simultaneously, although this would not apply to the single infant of a *P.verreauxi* troop. Ashmole (1962) points out that if there is clear advantage in the births occurring at one climatic season of the year, social synchrony would allow more accurate timing of births than if each individual relied on its own approximate timing.

In conclusion, the sexual behavior of *Lemur* presents many problems to the student of primate evolution. At least, however, it settles an old problem: continuous sexual attraction cannot be the cohesive force of primate society.

INFANTILE AND FRIENDLY RELATIONS

Washburn and DeVore (1961) propose as cohesive forces in a primate troop the interest the troop takes in the young infants and contact-grooming behavior. Hall and DeVore (1965) call contact and grooming in baboons friendly behavior. Andrew (1964a) relates adult grooming with lips and teeth to infant-suckling behavior in *Lemur*, *Galago*, and the Cercopithecoidea. It seems clear to me, at least in the Lemuroidea, that the adult friendly behavior originates directly from the contact, grooming, and play behavior of the infant (Plate 20).

Lemur and *Propithecus* infants both are carried by their mothers for the first few months of life. Baby *L.macaco* and *L.catta* cling to their mother's ventral surface with their tails over her waist like a safety belt. Adult *Lemur* often press their ventral surfaces against one another and, regardless of how they sit, sling their tails around the clump like feather boas. Infant *Propithecus* at first cling to their mother's stomach like young *Lemur* but soon transfer to her back and ride there for several months before becoming independent. *Propithecus* in a locomotive, then, are like a several-story mother-young group.

Grooming, like contact, begins with mother-infant relations. An infant *L.macaco* gives the same intense call when searching for the nipple as an adult gives in mutual genital licking (Andrew 1964a). In one hand-raised *L.macaco* the transition could be seen from suckling bare moist skin to licking such areas of a human being during grooming. This animal did not learn to use the lower incisors in grooming until placed at several months of age with other *Lemur*. A second hand-raised *L.macaco* never learned the full adult pattern of mutual genital grooming (R. Peck, personal communication). In both wild *P.verreauxi* and *L.catta* troop members other than its own mother attempted to groom the infant, and an infant *P.verreauxi* first reciprocated by grooming another animal at 6 weeks. Probably the use of the tooth-scraper is learned in infancy from the adults' grooming of the infant.

Play first appears in an infant when it is just beginning to leave its mother. In captive *L.macaco* and wild *P.verreauxi,* the first of all games is to leave the mother, crawl off two steps, then dash back, landing on the mother. Then the infant crawls two or three steps and dashes back, then three or four steps and back. If the mother sits still, the infant at last may move a meter away, usually upward, so that it drops down with some force and speed to its mother's fur. The whole procedure is an obvious game, from the slow adventure into the unknown to the delighted, reassuring bounce onto the mother's fur. At any movement from the mother, the infant scrambles toward her with frantic contact calls, legs and arms windmilling ineffectively.

This game begins with *P.verreauxi* so early that the hind legs do not have strength for leaps. The infant climbing a branch will first hop with both hind legs, like a climbing adult. Instead of moving the infant two or three body-lengths upward, the hop only lifts its rump away from the branch in a small bounce, while the hands remain in place. The infant then must haul itself upward hand over hand, as in climbing on fur. However, as the infant grows older, the game progresses, and soon the young juvenile can leap precisely in front of, or on top of, a troop member, and the clutching for fur becomes wrestling. In hand-raised *L.macaco* the game retains its early form even in adults, because a human being can double as mother and tree. Similarly, the hand-raised pet in the common contact play of wild *Lemur*—jump-on-and-wrestle—will soar for 20 feet through the air to thump onto a person's shoulder and then expect him to wrestle and groom.

If the components of friendly behavior, contact, grooming, and play, are directly derived from mother-infant relations, these two categories of behavior are somewhat interdependent, and their cohesive forces in primate society may be considered together.

In many primates, the entire troop takes a special interest in young infants. Primates vary, however, in the reactions of different ages and sexes.

Females and juveniles often crowd around a new mother and attempt to groom her and the infant. This is true in howlers (Carpenter 1934, 1965; Altmann 1959), langurs (Jay 1962, 1963, 1965), baboons (DeVore 1963*b*, Hall and DeVore 1965) and macaques (Southwick *et al.* 1965, Simonds 1965). Macaque and baboon females may even briefly steal infants (Rowell 1963*b*, Rowell, Hinde, and

Spencer-Booth, 1964; DeVore 1963*b*). This is not so true of chimpanzees and gorillas, but even there, older infants may form attachments to females besides their mother (Schaller 1963, 1965; Goodall 1965, Van Lawick-Goodall and Van Lawick 1965), while juveniles play with young babies (Van Lawick-Goodall and Van Lawick 1965). *Propithecus* and *L.catta* seem to resemble the others. *L.catta* offers a new variant of the general pattern, because a mother only permits other mothers to groom her newborn young.

The males of primate species differ even more in their attentions to infants (Mason 1965). Savannah baboon males are attracted to mothers and may groom their infants (Hall and DeVore 1965) or even adopt abandoned youngsters (Bolwig 1959). Marmoset males carry the young until they are half-grown, only transferring them to the mother for feeding (Fitzgerald 1935). Gibbon males play with their young (Carpenter 1940). Most other species are relatively aloof from infants (Southwick 1965; Simonds 1965; Jay 1965; Carpenter 1934, 1965; Goodall 1965), although male Japanese macaques occasionally "adopt" a juvenile (Itani 1959). *L.catta* males seemed to fall into a pattern of general indifference, but *P.verreauxi* males sought out and groomed infants as frequently as the females did. Except for marmosets, this makes *P.verreauxi* males more attentive than those of any species studied. The really interesting comparison is with the savannah baboons. It might be thought that the socialization of a baby *Propithecus* into its equable family group begins by contact with doting fathers, uncles, and brothers. However, the infant savannah baboon, groomed at first by the alpha male, will grow up to a troop life filled with overt and latent aggression, where a complex dominance hierarchy is the pattern of troop life. It is clear that the baboon males' association with infants protects them from predators, but we do not really know how socialization occurs.

There is one qualification. Although newborn infants seem to exert a strong binding force in many primate troops, this is not necessarily a year-round force. Many Cercopithecoidea breed seasonally (Lancaster and Lee 1965). *Lemur* and *Propithecus*, as stated, have an extremely sharp birth peak. Therefore we cannot expect the attraction of infants to hold the troop together from year to year.

The infant itself, however, learns social behavior from its early experiences. Some dramatic experiments on rhesus show that an infant raised with a pseudo-mother does not breed normally later or know how to rear its own offspring. Animals reared in isolation in cages, without even a pseudo-mother, are still more markedly abnormal (Harlow and Harlow 1961; Harlow, Harlow, and Hansen 1963).

Hand-raised *L.macaco* may be strongly imprinted on human

beings. One infant female that had human company from the age of 3 weeks to 5 months accepted a fur glove as a foster mother, clinging to it and crying when separated, but it preferred to have both human contact and the glove. The need for contact with a mother is certainly similar in *Lemur* and rhesus infants (Harlow 1958, Andrew 1964a). The hand-raised *L.macaco* was terrified of lemurs when reintroduced to them, but after several weeks accepted, and was accepted by, two subadults. At 2 years of age she bore an infant, which died at 3 weeks. Both she and another pet *Lemur* still groom human beings in preference to their mates or offspring. Clearly, then, the socialization of *Lemur* as well as of rhesus begins at a very early age.

Finally, the friendly relations are the most overt expression of cohesion in a primate troop. As with other aspects of behavior, their level varies enormously between genera: gorillas groom very little, baboons very often. *Cebus* grooms more than *Alouatta*, which grooms more than *Ateles*. Even related species may have different degrees of contact—bonnet macaques apparently more than pigtail macaques (Rosenblum, Kaufman, and Stynes 1964). In baboons and rhesus, grooming depends largely on rank, while in bonnets it has no relation to dominance (Hall and DeVore 1965, Southwick 1965, Simonds 1965). The form also varies: Old and New World monkeys part the fur, or pick up particles with their hands, while Lemuroidea simply grasp the fur with both hands while scraping with tongue and teeth. Still, contact and grooming between adults are two of the most widespread of primate characters. As Andrew (1964a) points out, some form of mutual grooming has been developed in most social mammals, but in no others is mutual grooming so common in ordinary social encounters. Play between adults, perhaps any play, also seems commoner to me in primates than other animals—also the tendency to turn normal greetings, grooming, or climbing into active locomotor and social play. The friendly relations are a group of relations in which primates elaborate their need for contact and their exuberant variation of pattern into social interactions between animals.

These friendly relations may be intensified in particular situations: at some phases of the female cycle (Michael and Herbert 1963, Rowell 1963a), in precopulatory behavior, and, of course, in mother- or troop-infant behavior. However, they persist throughout the year between many troop members in many circumstances. They can only be grouped as social interactions, not specifically sexual or maternal ones.

If the friendly relations are derived from mother-infant rela-

tions, it suggests that social primates retain many infantile characters. They are dependent on their fellows, and this need for a fellow's
company is expressed in infantile contact, grooming, and social play.
The *original* cohesive force in primate social evolution would then be
an infantile or juvenile attraction to others that was retained in the
adult.

Even if this is true, however, it must remain only a simplifying
theory. Infantile dependence is an idea at the level of a large-scale
behavioral drive, with all its conceptual difficulties (Hinde 1959). As
components of infantile behavior appeared with increasing frequency in the adult repertoire, they were assimilated to the adults'
system of behavior. As the adults became tolerant, then friendly to
each other, their greeting, contact, and grooming was no longer
"infantile" but "friendly."

INTELLIGENCE AND DEPENDENCE

Intelligence is a dangerous word and harder to define even than
"learning" or "insight." (Thorpe 1956). Nonetheless, higher primates
are, without any dispute, more intelligent than other mammals. Primates excel in perceiving new relations and in varying their responses
appropriately.

There are two groups of relations in which primates show their
ability: the first is relations between inanimate objects; the second,
between social fellows or social signals.

Psychologists usually measure a primate's learning ability or its
capacity for insight by tests with inanimate objects. Both chimpanzees (Köhler 1917) and *Cebus* (Harlow 1951) will invent and use
simple tools. Almost any primate will play and occupy itself with an
unfamiliar object (Harlow 1950, Jolly, 1964a, b). Most primates,
even marmosets, can be trained to learn discriminations between
objects, even to the extent of forming learning sets (Harlow 1949,
Miles and Meyer 1956). Rhesus achieve symbolic learning (Weinstein 1945) as do chimpanzees (Hayes 1951, Ferster 1964).

In this scale of intelligence, *Lemur* rank low. They score far
below both Old and New World monkeys on every type of problem,
from the simplest of insight tests in object manipulation to object
discrimination and delayed-response learning (Andrew 1962a, Jolly
1964b). *P.verreauxi* may be more attentive to the detail of objects
than *Lemur* (Andrew, personal communication), and seems more
visually alert in the wild. However, my impression is that *Propithecus*
is nearer to *Lemur* in this respect than to the Anthropoidea.

Captive *Lemur* do play with strange objects, but in one set of tests when they played with a hardware puzzle for 10 days, they had not learned to open it efficiently for a reward (Jolly 1964b).

This sort of play, and indeed the use of this sort of intelligence, is largely an artifact of captivity—the extraordinary richness of a human environment, or the barrenness of a cage. Examples of concern with inanimate objects in the wild are few enough to count.

Chimpanzees so far are the only primates known to make and use tools in the wild; Goodall in fact filmed (1964) baboons sitting around a party of chimpanzees fishing for termites, eager for the termites, but making no attempt to imitate the chimpanzees' technique. I watched *Cebus* show great ingenuity in their methods of detaching dead branches to drop on primatologists, while organgutans actually swing or half throw sticks toward an observer (Schaller 1963). The great apes build nests (as do many prosimians), which may be partly innate in both chimpanzees and gorillas (Bernstein 1962, Schaller 1963). The great apes, again, use leaves and vines for adornment in play.

There are no reliable records, though, of play with, or use of, inanimate objects in less advanced species, unless we count juvenile macaques' willingness to taste new food (Kawamura 1962) or locomotor play with springy branches and vines. Apparently most primates ignore inedible parts of the inanimate world, except under duress.

A different use of intelligence is in social relations. As yet we have no standard measures of social intelligence. Altmann (1965) has developed a statistical treatment of the chains of gesture and countergesture in rhesus macaques. He shows that some sequences are more probable than others—a result we would expect. The method allows calculation of the *length* of related sequences, which is a measure of the complexity of rhesus interaction. Perhaps lemurs will prove to have shorter chains of interaction than higher primates.

Another sign of complexity is the number of animals involved in an interaction. Baboon males may give "protected threats" (Kummer 1957); that is a weak animal challenges a superior, while standing in front of one still more dominant. The challenged cannot attack without himself being attacked. I did not see protected threats in *L.catta* troop 1, although if a dominant male saw a spat between two others, he would often threaten the winner. Similarly, troop 1 males had no "friends" whom they would support as part of a central hierarchy (Hall and DeVore 1965). Some baboon troops lack such a central hierarchy as well and have a linear dominance order like *L.catta* troop 1. It seems possible, though, that lemur aggressive interactions are less subtle than those of many monkeys.

Finally, we are learning that mother-infant relations in macaques may not be lost among general troop relations, but may lead to enduring kinship groups among the adults of a troop (Yamada 1963, Koford 1965). Relatives sit together, feed together in tolerance, and even pass on dominance status, so that the son of a powerful female may himself be dominant. Although the kinship grouping may be fundamental to primate social grouping, and hence be widespread among social primates, it adds still another dimension to primate study. Relations between animals would be affected not only by sex and age class, by dominance status, and by individual temperament, but also by kinship. There is as yet no data on lemurs (or many other primates) to reveal the role of kinship relations.

Thus, on further study, it may appear that lemur social relations are less sophisticated than those of the higher primates. However, it is clear that in many ways social lemurs resemble the higher primates. They, too, have permanent troops of all ages and sexes. They, too, give chains of gesture and countergesture, with animals responding not only to the particular signal, but to metacommunications about the serious or playful intention. They, too, recognize and respond to the individuals of the troop, adjusting their behavior for each animal and predicting future behavior of troop members. The Lemuroidea have developed a social intelligence in many ways comparable to that of monkeys, although their understanding of objects lags far behind.

Increasing social organization, through time, involves increasing learning and memory, for the rank and the idiosyncrasies of each troop member must be learned as well as the possible variations of behavior. Innate components determine less and less of the direction of behavior. But at the same time that primate social behavior demands learning, it makes it possible. Young animals are protected for a long time, and thus they are able to mature slowly while learning the ways of the troop. In many senses, a social animal is protected throughout its life, because the troop watches for and even challenges predators and raises intraspecific competition for food and space from the individual level to that of the group. Thus, there is a spiral interaction between intelligence and social dependence: intelligence, in primate societies, integrates an animal more surely with a group (where every animal plays a slightly different role) and makes the individual more dependent on the group while dependence allows and encourages ever-increasing social intelligence. (See Washburn and Hamburg 1965*b*, and Jolly 1966).

Lemur and *Propithecus* are both socially intelligent and socially dependent. They are, however, hopelessly stupid toward unknown inanimate objects. In this branch of the primates, the basic qualities of primate society have evolved without the formal inventive intelligence of true monkeys.

In another prosimian line, the Lorisoidea, *Galago senegalensis* lives in groups which may include more than one adult male (Haddow and Ellice 1964; Sauer and Sauer 1963) and *Galago crassicaudatus* may live in groups of up to nine animals. Contact with fur and mutual grooming play a large role in *Galago* maternal-infant behavior and other social interactions (Andrew 1964a, Buettner-Janusch 1964). *G.crassicaudatus* has a social defense mechanism: if one animal of a group screams, the others mob the predator with loud caws or even fling themselves on the predator in attack (personal observation). Among caged *Perodicticus potto,* each animal tends to sleep alone during the day, but a dominant male may regularly visit another male for half an hour's further sleep in early evening. During the night, caged pottos pay far more attention to each other than to available objects, except food (Cowgill 1964). Rahm (1960) likewise describes two strange pottos, which upon being introduced, clumped together like "true contact animals" (Hediger 1950). It seems that the tendencies toward social life in primates are more widespread, and hence older, than whatever led Oligocene monkeys to grow round skulls and brighter brains, leaving behind the long-nosed narrow-brained lemur. (Haddow, personal communication.)

That Oligocene step forward remains a mystery. I have suggested (Bishop 1962, Jolly 1964a) that a tendency to play with unfamiliar objects preceded the understanding of such objects. However, prosimians' tendency to play with objects, and monkeys' tendency to use them, rarely if ever finds expression in nature. It now seems likely to me that both the playfulness and the intelligence were advantageous largely in social contexts as a contact with fellows and as a means for predicting their behavior. General intelligence, which could be turned to new problems, not merely to social ones, would thus have been favored by selection for varied social understanding. (Washburn and Hamburg 1965b, and Jolly 1966.)

Social life, of course, does not *necessarily* lead to increased intelligence, as the lemurs themselves bear out. In both Australia and Madagascar it seems that it was competition from occupants of the same niches which evolved on intercommunicating land masses of the large continents that forced selection for increased intelligence. Also, "the presence of intelligent carnivores will result in the evolution of intelligent prey, and vice versa" (Andrew 1962a). "However, given the same level of such competition, a 'monkey' way of life led to higher intelligence than a 'mouse' way of life." (Andrew, personal communication.)

Intelligence and dependence interrelate, and reinforce each other. But if primates' social behavior derives from an Eocene ancestor, then dependence preceded primates' first great leap forward in intelligence. And if the prosimian, New World and Old World, primates evolved society separately, then in two branches there are clear examples of society without generally applicable intelligence: in the prosimians *Lemur* and *Propithecus*, probably in some Lorisoids, and in the New World marmosets and owl monkeys. I think we can guess that our own Old World ancestors probably passed through such a stage, when we had permanent society, without the cleverness even of the lowest round-brained monkey. Man's social emotions, then, are older than all his general ingenuity, and the first complexities of his mind grew with his social dependence.

Bibliography

Altmann, S. A. 1959. Field observations on a howling monkey society. *J. Mammal.* 40: 317–20.

Altmann, S. A. 1962. A field study of the sociobiology of rhesus monkeys, *Macaca mulatta*. *Ann. N.Y. Acad. Sci.* 102: 338–435.

Altmann, S. A. 1965. Sociobiology of rhesus monkeys. II. Stochastics of communication. *J. Theoret. Biol.* 8: 490–522.

Andrew, R. J. 1962a. Evolution of intelligence and vocal mimicking. *Science* 137: 585–89.

Andrew R. J. 1962b. The situations that evoke vocalization in primates. *Ann. N.Y. Acad. Sci.* 102: 296–315.

Andrew, R. J. 1963a. The origin and evolution of the calls and facial expressions of the primates. *Behaviour* 20: 1–109.

Andrew, R. J. 1963b. Trends apparent in the evolution of vocalization in the Old World monkeys and apes. *Zool. Soc. London, Symp.* 10: 89–101.

Andrew, R. J. 1964a. The displays of the primates. In *Evolutionary and genetic biology of the primates*, ed. J. Buettner-Janusch. Vol. 2, pp. 227–309. London and New York: Academic.

Andrew, R. J. 1964b. Vocalization in chicks, and the concept of "stimulus contrast." *Animal Behaviour* 12: 64–76.

Ashmole, N. P. 1962. The black noddy *Anous tenuirostris* on Ascension Island, Part I. General biology. *Ibis* 103b: 253–73.

Ashmole, N. P. 1963. The biology of the wideawake or sooty tern, *Sterna fuscata*, on Ascension Island. *Ibis* 103b: 297–364.

Attenborough, D. 1961. *Zoo quest to Madagascar*. London: Lutterworth.

Bastian, J. R. 1965. Primate signalling systems and human languages. In *Primate behavior*. Ed. I. DeVore, pp. 585–606. New York: Holt, Rinehart and Winston.

Bernstein, I. S. 1962. Response to nesting materials of wild born and captive born chimpanzees. *Animal Behaviour* 10: 1–6.

Bierens de Haan, J. A. 1930. Über das Suchen nach unsichtbaren Futter bei Affen und Halbaffen. Zugleich ein Beitrag zu der Frage nach dem Konkreten Verständnis dieser Tiere. *Z. Vergl. Physiol.* 11: 630–55.

Bierens de Haan, J. A. and Frima, M. J. 1930. Versuche über den Farbensinn der Lemuren. *Z. Vergl. Physiol.* 12: 603–31.

Bishop, A. 1962. Control of the hand in lower primates. *Ann. N.Y. Acad. Sci.* 102: 316–37.

Bishop, A. 1964. Use of the hand in lower primates. In *Evolutionary and genetic biology of the primates*, ed. J. Buettner-Janusch, Vol. 2, pp. 133–225. London and New York: Academic.

Bolwig, N. 1959. A study of the behavior of the Chacma baboon, *Papio ursinus*. *Behaviour* 14: 136–63.

Bolwig, N. 1960. A comparative study of the behavior of various lemurs. *Mém. Inst. Sci. Madagascar. Ser. A.* 14: 205–17.

Bourlière, F. 1964. *The natural history of mammals*. 3d ed. [trans. H. M. Parshley.] New York: Knopf.

Buettner-Janusch, J. 1962. Biochemical genetics of the primates—hemoglobins and transferrins. *Ann. N.Y. Acad. Sci.* 102: 235–48.

Buettner-Janusch, J. 1964. The breeding of galagos in captivity and some note on their behavior. *Folia Primatol.* 2: 93–110.

Buettner-Janusch, J. 1966. Origins of Man. New York: J. Wiley.

Buettner-Janusch, J. and Andrew, R. J. 1962. Use of the incisors by primates in grooming. *Am. J. Phys. Anthropol.* 20: 129–32.

Buettner-Janusch, J. and Buettner-Janusch, V. 1964. Hemoglobins of primates. In *Evolutionary and genetic biology of primates.* Ed. J. Buettner-Janusch, Vol. 2, pp. 75–91. New York and London: Academic.

Calhoun, J. B. 1963. *The ecology and sociology of the Norway Rat.* Bethesda, Md.: U.S. Dept. of Health, Education, and Welfare.

Carpenter, C. R. 1934. A field study of the behavior and social relations of the howling monkeys. (*Alouatta palliata.*) *Comp. Psychol. Monogr.* 10: 1–168.

Carpenter, C. R. 1940. A field study in Siam of the behavior and social relations of the gibbon (*Hylobates lar*). *Comp. Psychol. Monogr.* 16: 1–212.

Carpenter, C. R. 1962. Field studies of a primate population. In *Roots of behavior,* ed. E. L. Bliss, pp. 287–94. New York: Hoeber.

Carpenter, C. R. 1965. The howlers of Barro Colorado Island. In *Primate behavior,* ed. I. DeVore, pp. 250–91. New York: Holt, Rinehart, and Winston.

Chu, E. H. Y. and Bender, M. A. 1962. Cytogenetics of the primates and primate evolution. *Ann. N.Y. Acad. Sci.* 102: 253–66.

Clark, W. E. LeGros, 1959. *The antecedents of man.* Edinburgh: Edinburgh University Press.

Collias, N. and Southwick, C. A. 1952. A field study of population density and social organization in howling monkeys. *Proc. Am. Philos. Soc.* 96: 143–56.

Commissariat Général au Plan, 1962, *Economie Malgache, 1950–1960.* Tananarive.

Cowgill, U. M. 1964. Visiting in *Perodicticus. Science* 146: 1183–84.

Cowgill, U. M., Bishop, A. Andrew, R. J., and Hutchinson, G. E. 1962. An apparent lunar periodicity in the sexual cycle of certain prosimians. *Proc. Nat. Acad. Sci.* 48: 238–41.

Darwin, C. 1872. *Expression of the Emotions in Man and Animals.* London: J. Murray.

Decary, R. 1947. Epoques d' introduction des *Opuntia monacantha* dans le Sud de Madagascar. *Rev. Internat. de Botanique Appl. at d' Agr. Trop.* 27: 455–60.

DeVore, I. 1963a. A comparison of the ecology and behavior of monkeys and apes. Viking Fund Publ. in Anthropol. 37: 301–19.

DeVore, I. 1963b. Mother-infant relations in free-ranging baboons. In *Maternal behavior in mammals,* ed. H. L. Rheingold, pp. 305–35. New York: Wiley.

DeVore I. and Hall, K. R. L. 1965. Baboon ecology. In *Primate behavior,* ed. I DeVore, pp. 20–52. New York: Holt, Rinehart and Winston.

DeVore, I. and Washburn, S. L. 1963. Baboon ecology and human evolution. In *African ecology and human evolution,* eds. F. C. Howell and F. Bourlière, Vol. 36, pp. 335–53. Viking Fund Publ. in Anthropol.

Edwards, G. 1751. *Natural history of birds.* London.

Edwards, G. 1758. *Gleanings of natural history.* London.

Elliot-Smith, G. 1902. On the morphology of the brain in the mammalia, with special reference to that of the lemurs, recent and extinct. *Trans. Linn. Soc., London, Ser.* 2. 8: 319–432.

Elliot-Smith, G. 1927. *Essays in the evolution of man.* 2d ed. Oxford: Oxford University Press.

Ferster, C. B. 1964. Arithmetic behavior in chimpanzees. *Sci. Am.* 210: 98–107.

Fitzgerald, A. 1935. Rearing marmosets in captivity. *J. Mammal.* 16: 181–88.

Flacourt, E. de. 1661. *Histoire de la Grande Isle de Madagascar.* Paris: Pierre l'Amy.

Ford, C. E., Hamerton, J. L. and Sharman, G. B. 1957. Chromosome polymorphism in the common shrew. *Nature* 180: 392–93.

Gartlan, S., in press. The concept of dominance. *Proc. East Afr. Acad.*

Goodall, J. 1963. My life with wild chimpanzees. *Nat. Geogr.* 124: 272–308.

Goodall, J. 1965. Chimpanzees of the Gombe Stream Reserve. In *Primate behavior,* ed. I. DeVore, pp. 425–73. New York: Holt, Rinehart and Winston.

Goodman, M. 1962. Immunochemistry of the primates and primate evolution. *Ann. N.Y. Acad. Sci.* 102: 219–34.

Haddow, A. J. 1952. Field and laboratory studies of an African monkey, *Cercopithecus ascanius schmidti* Matschie. *Proc. Zool. Soc. London* 122: 297–394.

Haddow, A. J. and Ellice, J. M. 1964. Studies on bush-babies (*Galago spp.*) with special reference to yellow fever epidemiology. *Trans. Roy. Soc. Trop. Med. Hyg.* 58: 521–38.

Hall, K. R. L. 1962. Numerical data, maintenance activities, and locomotion of the wild chacma baboon (*Papio ursinus*). *Proc. Zool. Soc. London* 139: 181–220.

Hall, K. R. L. and DeVore, I. 1965. Baboon social behavior. In *Primate behavior,* ed. I. DeVore, pp. 53–110. New York: Holt, Rinehart, and Winston.

Harlow, H. F. 1949. The formation of learning sets. *Psychol. Rev.* 56: 51–65.

Harlow, H. F. 1950. Learning and satiation of response in an intrinsically motivated complex puzzle performance by monkeys. *J. Comp. Physiol. Psychol.* 43: 289–94.

Harlow, H. F., 1951. Primate learning. In *Comparative psychology,* ed. C. P. Stone, 3d ed., pp. 183–238. New York: Prentice-Hall.

Harlow, H. F. 1958. The nature of love. *Am. Psychol.* 13: 673–85.

Harlow, H. F. and Harlow, M. K. 1961. A study of animal affection. *Nat. Hist.* 60: 48–55.

Harlow, H. F., Harlow, M. K. and Hansen, E. W. 1963. The maternal affectional system of rhesus monkeys. In *Maternal behavior in mammals,* ed. H. L. Rheingold, pp. 254–81. New York: Wiley.

Harrison, J. L. 1952. Breeding rhythms of Selangor rodents. *Bull. Raffles Mus.* 24: 109–31.

Harrison, J. L. 1954. The moonlight effect on rat breeding. *Bull. Raffles Mus.* 25: 166–70.

Hayes, C. 1951. *The ape in our house.* New York: Harper.

Hediger, H. 1950. *Wild animals in captivity.* [trans. G. Sircom.] London: Butterworth.

Hill, W. C. O. 1953. *Primates: comparative anatomy and taxonomy.* I. *Strepsirhini.* Edinburgh: Edinburgh University Press.

Hill, W. C. O. 1955. *Primates: comparative anatomy and taxonomy.* 2. *Haplorhini: Tarsioidea.* Edinburgh: Edinburgh University Press.

Hill, W. C. O. 1957. *Primates: comparative anatomy and taxonomy.* 3. *Pithecoidea: Platyrhini. Families Hapalidae and Callimiconidae.* Edinburgh: Edinburgh University Press.

Hinde, R. A. 1959. Unitary drives. *Animal Behaviour* 7: 130–41.

Hinde, R. A. and Rowell, T. E. 1962. Postures and expressions in the rhesus monkey (*M. mulatta*). *Proc. Zool. Soc. London* 138: 1–32.

Humbert, H. 1947. Changements survenus dans la végétation du Sud de Madagascar. *Rév. Int. de Bot. Appl. et d' Agr. Trop.* 27: 441–44.

Humbert, H. 1955. Les territoires phytogéographiques de Madagascar. Leur cartographie. *Année biologique* 31: 195–204.

Hutchinson, G. E., 1957. Concluding remarks. *Cold Spring Harbor Symp. Quant. Biol.* 22: 415–27.

Hutchinson, G. E. 1959. Homage to Santa Rosalia, or why are there so many kinds of animals? *Am. Nat.* 93: 145–59.

Hutchinson, G. E. 1965. *The ecological theater and the evolutionary play.* New Haven and London: Yale University Press.

Hutchinson, G. E. and MacArthur, R. H. 1959. A theoretical ecological model of size distributions among species of animals. Am. Nat. 93: 117–25.

Huxley, T. H. 1861. On the brain of *Ateles paniscus. Proc. Zool. Soc. London.* 247–60.

Huxley, T. H. 1863. *Man's place in nature.* New York: Macmillan and Co.

Itani, J. 1959. Paternal care in the wild Japanese monkey, *Macaca fuscata fuscata. Primates* 2: 61–93.

Itani, J. 1963. Vocal communication in the wild Japanese monkey. *Primates* 4: 11–66.

Jay, P. 1962. Aspects of maternal behavior among langurs. *Ann. N.Y. Acad. Sci.* 102: 468–76.

Jay, P. 1963. The Indian langur monkey (*Presbytis entellus*). In *Primate social behavior,* ed. C. H. Southwick, pp. 114–23. Princeton: Van Nostrand.

Jay, P. 1965. The common langur of north India. In *Primate behavior,* ed. I. DeVore, pp. 197–249. New York: Holt, Rinehart and Winston.

Jolly, A., in press. Sex ratio and dominance in social lemurs. *Proc. East Afr. Acad.*

Jolly, A. 1964a. Prosimians' manipulation of simple object problems. *Animal Behaviour.* 12: 560–70.

Jolly, A. 1964b. Choice of cue in prosimian learning. *Animal Behaviour.* 12: 571–77.

Jolly, A. 1966. Lemur social behavior and primate intelligence. *Science* 153 (3735): 501–6.

Kawamura, S. 1962. The process of sub-culture propagation among Japanese macaques. *Primates* 2: 43–60.

Kent, R. K. 1962. From Madagascar to the Malagasy Republic. New York: Praeger.

Klüver, H. 1933. *Behavioral mechanisms in monkeys.* Chicago: University of Chicago Press.

Köhler, W. 1917. *The mentality of apes.* [trans. Ella Winter.] New York: Harcourt Brace & Co.; London: K. Paul, Trench, Trübner.

Koford, C. B. 1965. Population dynamics of rhesus monkeys on Cayo Santiago. In *Primate behavior,* ed. I. DeVore, pp. 160–74. New York: Holt, Rinehart, and Winston.

Kummer, H. 1957. Sozialesverhalten einer Mantelpavien–Gruppe. *Sweiz. Z. Psychol.,* No. 33.

Kummer, H. and Kurt, F. 1963. Social units of a free-living population of hamadryas baboons. *Folia Primatol.* 1: 4–19.

Lancaster, J. B. and Lee, R. B. 1965. The annual reproductive cycle in monkeys and apes. In *Primate behavior,* ed., I. DeVore, pp. 486–513. New York: Holt, Rinehart, and Winston.

Langford, J. B. 1963. Breeding behaviour of *Hapale jacchus* (common marmoset). *S. Afr. J. Sci.* 59: 299–300.

Linnaeus, C. V. 1758. *Systema naturae.* Stockholm: Holmiae.

MacArthur, R. and Levins, R. 1964. Competition, habitat selection and character displacement in a patchy environment. *Proc. Nat. Acad. Sci.* 51: 1207–10.

Marler, P. 1965. Communication in monkeys and apes. In *Primate Behavior,* ed., I. DeVore, pp. 544–84. New York: Holt, Rinehart and Winston.

Mason, W. A. 1965. The social development of monkeys and apes. In *Primate Behavior*, ed. I. DeVore, pp. 514–43. New York: Holt, Rinehart, and Winston.

Mayr, E. 1963. *Animal species and evolution*. Cambridge, Mass.: Harvard University Press.

Menaker, W. and Menaker, A. 1959. Lunar periodicity and human reproduction: a likely unit of biological time. *Am. J. Obstet. Gynecol.* 77: 905–14.

Meylan, A. 1960. Contribution à l'étude du polymorphisme chromosomique chez *Sorex araneus* L. (Mamm. Insectivora). Note préliminaire. *Rév. Suisse Zool.* 67: 258–61.

Michael, R. P. and Herbert, J. 1963. Menstrual cycle influences grooming behavior and sexual activity in the rhesus monkey. *Science* 140: 500–501.

Miles, R. C. and Meyer, D. R. 1956. Learning sets in marmosets. *J. Comp. Physiol. Psychol.* 49: 219–22.

Millot, J. 1952. La faune malgache et le mythe Gondwanien. *Mém. Inst. Sci. Madagascar Sér. A.* 7: 1–33.

Milne-Edwards, A. and Grandidier, A. 1875, 1890–1896. *Histoire naturelle des mammifères: Histoire physique, naturelle, et politique de Madagascar.* Vols. VI, IX, X. Paris.

Mohr, C. O. 1947. Table of equivalent populations of North American small mammals. *Am. Midl. Nat.* 37: 223–49.

Montagna, W. 1962. The skin of lemurs. *Ann. N.Y. Acad. Sci.* 102: 190–209.

Montagna, W. R., Yasuda, K., and Ellis, R. A. 1961. The skin of primates. V. The skin of the black lemur (*Lemur macaco*). *Am. J. Phys. Anthropol.* 19: 115–30.

Moog, G. 1959. Geburt eines bastardes Mohrenmaki x Schwarzkopfmaki im Zoologische Garten Saarbrücken. *Der Zoologische Garten* 25: 99–104.

Morris, D. 1956. The feather postures of birds and the problem of the origin of social signals. *Behaviour* 9: 75–113.

Parkes, A. S. (ed.) 1956. *Marshall's physiology of reproduction.* London: Longmans.

Parkes, A. S. and Bruce, H. M. 1961. Olfactory stimuli in mammalian reproduction. *Science* 134: 1049–54.

Petter, J. J., 1962a. Recherches sur l' écologie et l' éthologie des Lémuriens malgaches, *Mém. du Mus. Nat. de l' Hist. Naturelle. Sér. A.* 27: 1–146.

Petter, J. J., 1962b. Ecological and behavioral studies of Madagascar lemurs in the field. *Ann. N.Y. Acad. Sci.* 102: 267–81.

Petter, J. J. 1962c. Ecologie et éthologie comparées des Lémuriens malgaches. *La Terre et la Vie* 109: 394–416.

Petter, J. J. 1965. The lemurs of Madagascar. In *Primate behavior.* Ed. I. DeVore, pp. 292–319. New York: Holt, Rinehart, and Winston.

Petter-Rousseaux, A. 1962. Recherches sur la biologie de la réproduction des primates inférieurs. *Mammalia* 26 (Suppl. 1): 1–88.

Petter-Rousseaux, A. 1964. Reproductive physiology and behavior of the Lemuroidea. In *Evolutionary and genetic biology of the primates*, ed. J. Buettner-Janusch. Vol. 2. New York and London: Academic.

Pocock, R. I. 1918. On the external characters of the lemurs and of *Tarsius. Proc. Zool. Soc. London*, 19–53.

Polo, M. 1295 (1875). *The book of Ser Marco Polo the Venetian concerning the kingdoms and marvels of the East.* [trans. H. Yule.] London: John Murray.

Rahm, U. 1960. Quelques notes sur le potto de Bosman. *Bull. Inst. Français de l' Afr. Noire.* 22: 331.

Rand, A. L. 1935. On the habits of some Madagascar mammals. *J. Mammal.* 16: 89–104.

Rand, A. L. 1936. The distribution and habits of Madagascar birds. *Bull. Am. Mus. Nat. Hist.* 72: 143–499.

Rosenblum, L. A., Kaufman, I. C., and Stynes, A. J. 1964. Individual distance in two species of macaque. *Animal Behavior* 12: 338–42.

Rowell, T. E. 1963a. Behavior and female reproductive cycles of rhesus macaques. *J. Reprod. Fertil.* 6: 193–203.

Rowell, T. E., 1963b. The social development of some rhesus monkeys. In *Determinants of infant behavior,* ed. B. M. Foss, pp. 35–49. London: Methuen.

Rowell, T. E., in press. Ecology of Uganda forest-living baboons. *Proc. East Afr. Acad.*

Rowell, T. E. and Hinde, R. A. 1962. Vocal communication by the rhesus monkey (*Macaca mulatta*). *Proc. Zool. Soc. London* 138: 279–94.

Rowell, T. E., Hinde, R. A., and Spencer-Booth, Y. 1964. "Aunt"-infant interaction in captive rhesus monkeys. *Animal Behaviour* 12: 219–26.

Sade, D. St. 1964. Seasonal cycle in size of testes of free-ranging *Macaca mulatta. Folia Primatol.* 2: 171–80.

Sauer, G. F. and Sauer, E. M. 1963. The South-West African bushbaby of the *Galago senegalensis* group. *J. South West Afr. Sci. Soc.,* 16: 5–36.

Schaller, G. 1963. *The mountain gorilla: ecology and behavior.* Chicago: University of Chicago Press.

Schaller, G. 1965. The behavior of the mountain gorilla. In *Primate behavior,* ed. I. DeVore, pp. 324–67. New York: Holt, Rinehart, and Winston.

Shaw, G. A. 1879. A few notes upon four species of lemurs, specimens of which were brought alive to England in 1878. *Proc. Zool. Soc. London* 132–36.

Simons, E. L. 1962. Fossil evidence relating to the early evolution of primate behavior. *Ann. N.Y. Acad. Sci.* 102: 282–93.

Simons, E. L. 1963. A critical reappraisal of the Tertiary primates. In *Evolutionary and genetic biology of the primates,* ed. J. Buettner-Janusch, Vol. 1, pp. 65–125. New York and London: Academic.

Simons, E. L. 1964. The early relatives of man. *Sci. Am.* 211: 50–62.

Simonds, P. B. 1965. The bonnet macaque in south India. In *Primate behavior,* ed. I. DeVore, pp. 175–96. New York: Holt, Rinehart, and Winston.

Simpson, G. G. 1945. The principles of classification and a classification of the mammals. *Bull. Am. Museum Nat. Hist.* 85: 1–350.

Southwick, C. 1962. Patterns of intergroup social behavior in primates, with special reference to rhesus and howling monkeys. *Ann. N.Y. Acad. Sci.* 102: 436–54.

Southwick, C. H., Beg, M. A., and Siddiqi, M. R. 1965. Rhesus monkeys in north India. In *Primate behavior,* ed. I. DeVore, pp. 111–59. New York: Holt, Rinehart, and Winston.

Sparks, J. H. 1964. Flock structure of the red avadavat with particular reference to clumping and allopreening. *Animal Behaviour* 12: 125–36.

Starnmühlner, F. 1960. Beobachten am mausmaki (*Microcebus murinus*). *Natur. Volk* 90: 194–204.

Struhsacker, T., in press. Behavior of vervets (*Cercopithecus aethiops*) *Proc. East Afr. Acad.*

Sugiyama, Y. 1964. Group composition, population density, and some sociological observations of Hanuman langurs (*Presbytis entellus*) *Primates* 5: 7–38.

Thorpe, W. H. 1956. *Learning and instinct in animals.* London: Methuen.

Van Lawick–Goodall, J. and Van Lawick, H. 1965. New discoveries among Africa's chimpanzees. *Nat. Geogr.* 128: 802–31.

Washburn, S. L. and DeVore, I. 1961. The social life of baboons. *Sci. Am.* 204: 62–71.

Washburn, S. and Hamburg, D. A. 1965a. The study of primate behavior. In *Primate behavior,* ed. I. DeVore, pp. 1–15. New York: Holt, Rinehart, and Winston.

Washburn, S. L. and Hamburg, D. A. 1965b. The implications of primate research. In *Primate behavior,* ed. I. DeVore, pp. 607–22. New York: Holt, Rinehart, and Winston.

Weinstein, B. 1945. The evolution of intelligent behavior in rhesus monkeys. *Genet. Psychol. Monogr.* 31: 3–48.

Wight, H. M. 1931. Reproduction in the eastern skunk (*Mephitis mephitis nigra.*) *J. Mammal.* 12: 42–47.

Wynne-Edwards, V. C. 1962. *Animal dispersion in relation to social behavior.* Edinburgh. Oliver and Boyd.

Yamada, M. 1963. A study of blood-relationship in the natural society of the Japanese macaque. *Primates* 4: 43–66.

Zuckerman, S. 1932. *The social life of monkeys and apes.* London: K. Paul, Trench, Trübner.

Zuckerman, S. 1933. *Functional affinities of man, monkeys, and apes.* London: K. Paul, Trench, Trübner.

Author Index

Subject Index

NOTE:
The three species described in this volume are abbreviated as follows in the index:

Lemur catta—L.c.
Lemur macaco—L.m.
Propithecus verreauxi verreauxi—P.v.v.

Activity
  agonistic
    comparison of *L.c.* and *L.m.*, 155
    L.c., 74, 83, 95, 100–101, 103–10,
      113, 132
    L.m., 122, 123
  General
    L.c., 44, 72, 77, 93, 95, 103
    P.v.v., 24, 29, 33–34, 44, 55–58,
      77, 93, 94
  Play
    L.c., 95, 160, 161
    L.m., 161
    P.v.v., 58–59, 160, 161
  Sexual
    L.c., 80, 88–89, 90, 95
    L.m., 122, 123, 126, 128–29
    P.v.v., 21, 23, 28
  Submissiveness
    L.c., 115
  by time
    lemuroidea, 4 Table I-2
    diurnal
      L.c., 81, 133
      L.m., 133
      P.v.v., 133
    nocturnal
      L.c., 79–80
Agapornis cana and *P.v.v.*, 42. *See
  also* Relations with other species
  (birds and bats)
Aggression, as emotion, 1, 60–62. *See
  also* Activity (agonistic); Emo-
  tions (aggression)
Agonistic activity. *See* Activity (ago-
  nistic); Emotions (aggression)
Anatomy
  L.c., 99, 103, 149
  L.m., 119, 146
  Lemuroidea, 36
  P.v.v., 36, 64
Apes
  and man, 1
  and monkeys, 1
  and olfactory communication, 132

Arabs and Madagascar. *See* Madagascar
Avahi and Indri, 20
Aye-aye
  displays, vocal, 142
  taxonomy, 3
  see also Lemuroidea

Baboons
  and aggression, 62, 110, 154
  and grins, 130
  intelligence, 164
  mutual grooming, 160–61
  play, 160–61
  and *P.v.v.*, 62
  see also Relations with other species
    (baboons)
Beetles. *See* Relations with other species (beetles)
Behavior
  aggression, 134, 135, 153
  play, 134
  submission, 135
  see also Activity; Emotions
Brachiators, anatomy, comparison
  with *P.v.v.*, 35
Brain size of lemurs and primates, 2,
  8, 166. *See also* Monkey
Branch height
  L.c., 28–29, 73 74, 150–51
  L.m., 121, 150–52
  P.v.v., 28–29, 73–74, 150–52
  and scent-marking, *L.c.*, 101
"Blaze Nose Troop, *P.v.v.*, 23, 24, 46
  Table II-6, 64, 66, 68

Care for the young, 4 Table I-2
  as emotion, 1
  see also Child rearing; Lemuroidea
Cats. *See* Relations with other species
  (cats)
Ceboidea. See Monkeys (New World)
Cebus, intelligence, 163, 164
Cercopithecoidea. See Monkeys (Old
  World)

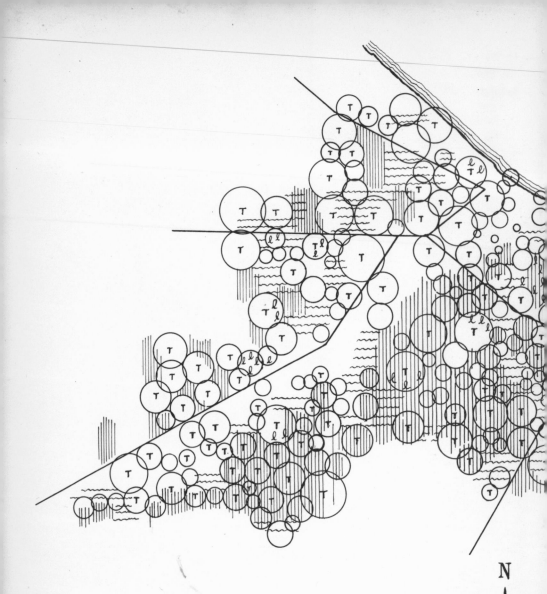

STUDY AREA

SCALE: |———|———|———|———| 100 m

——— PATHS

Ⓣ TAMARIND TREES

◯ OTHER LARGE TREES

|||||| TREES LESS THAN 13 M TALL

〰〰 BUSHES

ℓ ℓ ℓ LIANAS

Blank BARE GROUND OR MEADOW

N